THE POLLINATOR VICTORY GARDEN

Win the War on Pollinator Decline with Ecological Gardening

HOW TO ATTRACT AND SUPPORT BEES, BEETLES, BUTTERFLIES, BATS, AND OTHER POLLINATORS

Kim Eierman

Brimming with creative inspiration, how-to projects, and useful information to enrich your everyday life, Quarto Knows is a favorite destination for those pursuing their interests and passions. Visit our site and dig deeper with our books into your area of interest: Quarto Creates, Quarto Cooks, Quarto Homes, Quarto Lives, Quarto Drives, Quarto Explores, Quarto Gifts, or Quarto Kids.

First Published in 2020 by Quarry Books, an imprint of The Quarto Group,
100 Cummings Center, Suite 265-D, Beverly, MA 01915, USA.
T (978) 282-9590 F (978) 283-2742 QuartoKnows.com

Quarry Books titles are also available at discount for retail, wholesale, promotional, and bulk purchase. For details, contact the Special Sales Manager by email at specialsales@quarto.com or by mail at The Quarto Group, Attn: Special Sales Manager, 100 Cummings Center, Suite 265-D, Beverly, MA 01915, USA.

10 9 8 7 6 5 4 3 2

ISBN: 978-1-63159-750-3

Digital edition published in 2020
eISBN: 978-1-63159-751-0

Library of Congress Cataloging-in-Publication Data

Eierman, Kim, author.

The Pollinator victory garden : win the war on pollinator decline with ecological gardening : how to attract and support bees, beetles, butterflies, bats, and other pollinators / Kim Eierman.

LCCN 2019025269 | ISBN 9781631597503 (trade paperback)

1. Pollination. 2. Pollinators. 3. Gardening to attract wildlife.

LCC QK926 .E34 2020 | DDC 576.8/75--dc23

Design & Layout: Tanya Jacobson, crsld.co

Front Cover Image: Kim Eierman. Native bumble bee nectaring on goldenrod.

Back Cover Image: Carolyn Summers. Pollinator garden at The Native Plant Center, Westchester County, New York

Photography: Kim Eierman (pages 6, 12, 14 [left], 21, 33, 36, 39, 47, 54 [right], 58, 72, 73, 89, 92, 94, 101, 117, 124, 125, 126, 128, 130, 131, 135, 137, 142), Heather Holm (pages 11, 16, 17, 18, 19, 22 [left], 23, 29, 31, 37, 40, 45, 74, 80, 93, 95, 96, 98, 103, 104, 107, 110, 112, 114, 140, 146), Carolyn Summers (pages 4, 10, 14 [right], 20, 26, 28, 32, 34, 44, 46, 48, 51, 52, 56, 61, 62, 65, 66, 67, 69, 70, 77, 78, 83, 85, 86, 102, 108, 122, 123, 127, 132, 136, 138, 142, 143, 144), Shutterstock (pages 22 [right], 38, 53, 54 [left], 116, 118)

Illustration: Greta Moore

Printed in China

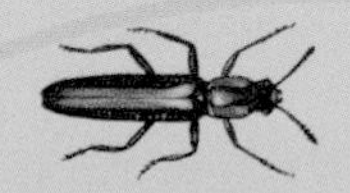

In memory of Rachel Carson, author of the seminal book *Silent Spring*, who warned us a half a century ago about the dangers of synthetic pesticides.

Thank you to Heather Holm and Carolyn Summers, my much appreciated collaborating photographers. Your wonderful photos help underscore the beauty and importance of pollinators, as do your books, your design work, and your environmental efforts.

Pollinator garden at The Native Plant Center, Westchester County, New York

TABLE *of* CONTENTS

INTRODUCTION

Pollinators are critical to our food supply and responsible for the pollination of the vast majority of all flowering plants on our planet. But many pollinators are in trouble, and the reality is that most of our landscapes have little to offer them. I wrote this book to help you change that. You *can* create a beautiful landscape that attracts and supports many different species of pollinators.

The Many Types of Pollinators

Animal pollinators include not just bees but also many other types of pollinators, including insects and mammals. Beetles, bats, birds, butterflies, moths, flies, wasps, and even some mosquito species can be pollinators. This is not to say that all species within these various groups are pollinators, but within each group many species are. Beetles are the largest and most ancient group of pollinators on Earth, and likely one of the groups we rarely garden for. It's time to start thinking about supporting a variety of pollinators in our landscapes and gardens.

Pollinator Decline

Pollinator decline is an ecological reality of our time, well publicized in the media and evident to anyone who gardens or even steps outside during the growing season. It is not your imagination—research studies confirm what we are seeing, or, rather, *not* seeing. It is fair to say that we are facing a pollinator crisis. It has been a long time in the making, but many of us simply have not noticed until recently.

Worldwide, there are an estimated 20,000 species of bees, with more than 3,600 known species in the United States and Canada, but many bee species are suffering losses. In 2017, the Center for Biological Diversity published disturbing research findings on native bees in North America and Hawaii. Of those species with sufficient data to study and assess, *more than half* of those native bee species were declining. In addition, one in four native bee species in North America and Hawaii is imperiled and at an increasing risk of extinction. And that is just native bees; European honey bees have been facing their own significant threats in recent years.

Major Causes of Pollinator Decline

There are numerous threats to pollinators in our environment, including the dwindling of resources they need to exist. Ongoing construction and development rob pollinators of precious habitat and flowers. New housing developments, corporate parks, shopping malls, and even home renovations and additions displace pollinators from the habitats they need. Restoring native habitat has rarely been a primary consideration for most homeowners, but it should be. I'll give you the knowledge you need in order to change that.

Residential, commercial, and municipal landscapes are filled with vast green pollinator deserts—better known as lawns. These vast expanses are ecological wastelands for bees and other pollinators, save for the few lawns that are allowed to contain pollinator-friendly blooming "weeds," such as clovers and dandelions. Imagine all of the pollinators that once flourished on these sites before the lawns were created. By planting just a little bit differently and by tweaking your landscape aesthetic, you can transition your landscape into a pollinator haven.

Widespread pesticide use is another cause of pollinator decline. Pollinators can be exquisitely sensitive to man-made pesticides, but even some organic pesticides can be fatal to delicate creatures like bees. By adopting garden management strategies that do not require pesticides, you can keep pollinators, and humans, healthier. By using proper cultural practices in your garden, by choosing the right plant for the right location, and by attracting "nature's pest control," beneficial insects that act as natural enemies, you can keep nature in balance and give pollinators a fighting chance.

Climate change is another threat to pollinators, often throwing plant activity such as bloom time out of sync with pollinator activity. Thoughtful gardening strategies can help fill some of these gaps. Planting for an overlapping succession of bloom in the garden throughout the entire growing season has become critically important in the face of climate change.

The Pollinator Victory Garden

During World War I, a movement was started to protect food supplies. In the United States, Canada, the United Kingdom, and Australia, vegetable, fruit, and herb gardens were planted in residential landscapes, community gardens, and public parks. Known as food gardens for defense, these plots were promoted as Victory Gardens. In addition to providing a reliable food source, Victory Gardens were encouraged as a way to boost civilian morale and empower participants to help the war effort off the battlefield. Two decades later, when World War II began, Victory Gardens were popularized once again. The United States Department of Agriculture estimates that more than 20 million Victory Gardens were planted in the United States during World War II.

The passion and urgency that inspired those Victory Gardens long ago is needed today to meet another threat to our food supply and our environment—the steep decline of pollinators. *Now* is the time for a new gardening movement. Every yard, garden, rooftop, porch, patio, and corporate landscape can help win the war against pollinator decline.

You don't have to be an entomologist to realize that pollinators are in trouble, and you don't have to be a professional landscaper or horticulturist to do something about it. *The Pollinator Victory Garden* is for all of us who care about the environment, who want to make it better, and who enjoy having a beautiful landscape that is filled with life. This book shows you how to use ecological landscaping and native plants to benefit not just pollinators but also the overall health of the ecosystem that is your landscape.

I hope you enjoy the book, and most important, I hope you relish the wonderful results that come when you implement these tips.

More pollinators? Yes, please.

1 Essentials of Pollinators and Pollination

Pollination is something we take for granted, but it's essential to life on Earth. It is the means by which most flowering plants reproduce. As you'll recall from biology class, pollination involves the physical movement of pollen from the male part of a plant to the female part of a plant. Without pollination, a plant can't be fertilized so that seeds can form. Pollination can happen in several ways:

- by animals (including insects)
- by wind
- by water

Most flowering plants are animal pollinated, with insects being the largest group of pollinators. Plants such as grasses, conifers, and oaks depend on wind pollination. Water pollination typically occurs with flowering aquatic plants.

Native plants, such as butterfly weed (*Asclepias tuberosa*), attract many different pollinators.

The Importance of Animal Pollinators

Pollinators and most flowering plants have a mutually beneficial relationship. Pollinators visit flowers to drink nectar or eat pollen, depending on the type of pollinator. As they feed, they move grains of pollen from flower to flower on the same plant or from flower to flower on different plants. Pollination is not an intentional act by pollinators but usually a consequence of feeding. In some cases, pollen is the only food source of the pollinator; not all pollinators eat nectar. Depending on the plant species, its flowers may be visited by a number of different pollinator types or it may attract a very narrow group of pollinators.

At least 80 percent of all flowering plants on Earth depend on pollinators for reproduction. Pollinators are vitally important to most ecosystems, with some pollinators even functioning as a keystone species—that is, a species upon which other species depend in an ecosystem. If a keystone species becomes extinct or vanishes from an ecosystem in which it has evolved, that ecosystem fundamentally changes, and usually not for the better.

In addition to providing pollination services, pollinators are part of food webs (interdependent food chains) and can be a nutritious protein meal

Native squash bees pollinating a squash plant

for creatures farther up a food chain. While it may be painful to watch a bird eat a butterfly or bee, every creature must eat. Food webs are the basis of ecosystems, and pollinators are part of them. When you garden, consider that you are really planting for food webs, not only for specific species. Specialist pollinators do need extra help in our gardens though. These specialists have evolved with certain plants they depend upon and vice versa. Include those plants in your garden—for example, the hibiscus bee specializes on hibiscus.

The world's food supply is heavily dependent on pollinators—mainly bees. While not all food crops require pollination, according to the United Nations, 84 percent of crops grown for human consumption need pollinators to increase their yields and enhance their quality. While various insects pollinate food crops, bees—including but not limited to honey bees—are the most important pollinators for food crops. The United States Department of Agriculture estimates that 80 percent of insect-based crop pollination is performed by bees. Marla Spivak, a renowned entomologist and bee expert, explains if *bees* don't have enough to eat, *we* won't have enough to eat. Because many of our food crops are grown as giant monocultures and pollinators need a variety and abundance of food to eat, the more plant diversity we can bring into our gardens, the better.

The lawn—a pollinator desert

Threats to Pollinators

Pollinators face a magnitude of threats to their survival that seems disproportionate to their ecological value as creatures that are vital to life on Earth. Many of these challenges to pollinators result from human activity and decisions made without consideration of ecological consequences. Building and development takes its toll on pollinators as their habitat, nesting sites, and floral resources vanish. One of the worst threats is the fragmentation of pollinator habitat, which occurs when buildings, roads, or clear-cutting split up natural areas, dramatically reducing pollinator survival.

A decrease in biodiversity of plants, an emphasis on lawns, the negligible use of native plants, and the explosion of invasive plant species are all taking their toll on pollinators. Pesticide use in agriculture and in commercial and residential landscapes further stacks the deck against pollinator survival. Already weakened by these myriad challenges, pollinators become more susceptible to new pests and diseases that plague them. Research on pollinators from Penn State shows that the combination of exposure to pesticides and reduced nutrition may make bees more susceptible to the other major suspects in bee decline: parasites and pathogens. By decreasing the human-driven threats to pollinators, we can surely help them rebound.

Removing Threats

When creating your Pollinator Victory Garden, focus on increasing the resources pollinators need and improving their habitats. You can start quite simply by reducing your lawn (a pollinator wasteland) and replacing it with a variety of flowering native plants that you maintain free of pesticides. The accompanying chart details some of the impediments to pollinators (minuses) and the improvements you can make to help them (pluses).

Impediments to Pollinators (Avoid This)	Improvements for Pollinators (Do This)
Large lawn (the green desert)	Reduce or eliminate the lawn and replace with flowering native plants.
Landscapes with few flowers	Plant an abundance of flowering native trees, shrubs, and perennials.
Time periods without blooms	Design for a succession of blooming plants throughout the growing season.
Monoculture plantings	Include diverse plantings of many different flowering plant species that bloom at different times.
Mostly nonnative plants	Emphasize regional native plants that pollinators have evolved with.
Invasive plants	Remove invasive plants (even if pollinators like them) and replace with native plants.
Pesticide use	Eliminate pesticides; attract beneficial insects to keep insect pests in check.
Lack of nesting sites	Provide patches of bare soil, brush piles, dead trees and logs, plants with pithy stems, stone walls with crevices, and man-made bee houses.
Fragmented habitat	Join with neighbors to create pollinator corridors that run from one landscape to the next, and the next . . .

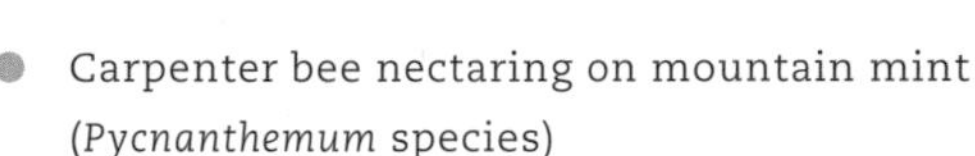

Carpenter bee nectaring on mountain mint (*Pycnanthemum* species)

Hummingbird moth nectaring on wild bergamot (*Monarda fistulosa*)

Types of Pollinators

Most animal pollinators are insects. While bees are arguably the best-known animal pollinators, beetles, butterflies, moths, flies, and wasps are other pollinators you will most likely attract and support in a Pollinator Victory Garden. You may also be able to entice nectar-eating birds (nectivores) such as hummingbirds and orioles, depending on your location. Hummingbirds are the largest group of pollinating birds in North America, with an estimated 350 species. Some bird pollinators even specialize in pollinating specific plants, such as the white wing dove, which is an important pollinator of the saguaro cactus in Arizona.

In tropical and desert areas, bats may be important pollinators to support in your landscape. More than 300 types of fruit, including mangoes, bananas, and guavas, depend on bat pollination. Even the saguaro cactus and agave plants are pollinated by bats.

Less common pollinators include certain species of mosquitoes, ants, and even some mammals and lizards, depending on your country and region. Which pollinators are important? All of them. If they have evolved in your locale, they are part of your ecosystem and they have ecological value.

Types of Pollinators

Insects—the largest category of animal pollinators.

- Bees — most important group for pollination of food crops
- Beetles — largest group of insect pollinators
- Butterflies — often incidental pollinators, but some are very important
- Moths — usually nocturnal, but some species are active during the day
- Flies — often bee mimics; identified by only one pair of wings
- Wasps — less efficient pollinators than bees; specialist wasps on figs, orchids
- Ants — nectar-loving but not very effective at pollination
- Mosquitoes — limited pollinators of certain native orchids

Birds—pollinating birds include hummingbirds, spiderhunters, sunbirds, honeycreepers, honeyeaters, and some parrots.

Mammals—pollinating mammals include some species of bats, marsupials, monkeys, lemurs, and rodents.

Lizards—pollinating lizards are found mostly on islands.

Bees and Beyond

It can be difficult to convince people to garden for pollinators besides bees. Actually, the fact that bees are even being considered in gardens means that we have made much progress in the last decade or so. Not long ago, butterflies, and perhaps hummingbirds, were the only pollinators that were truly welcome in a majority of landscapes. Bees were often viewed as the wards of beekeepers or inconvenient wild creatures to be avoided and feared for their sting, certainly not creatures to be encouraged in managed landscapes. Much of this bee fear was based on misinformation; most bees are quite docile and will usually sting only if they're threatened around their nesting area.

Now that home gardeners are starting to embrace bees, we need to expand our ecological horizons to garden for all types of pollinators. If we don't, we will lose even more species. Biodiversity is *the key* to healthy ecosystems and a critical tool in fighting climate change. A healthy garden ecosystem is one that is full of many living creatures, including pollinators. Plant diversity = animal diversity. In your garden, include a diversity of native plants, such as trees, shrubs, vines, and perennials, that have a multitude of shapes and sizes to attract a greater diversity of pollinators.

Open, accessible flower: pollinating fly on a native aster (*Symphyotrichum* species)

Pollinator Variations and Garden Considerations

Pollinators vary in many ways that impact what a pollinator garden must provide. The tongue length of a pollinator is a major factor in determining the flowers it can access. A long-tongued species of bee, butterfly, or hummingbird will be able to reach nectar in a long, tubular flower, while a short-tongued pollinator will need flatter, more open, accessible flowers. By planting flowering plants with various flower forms, you will attract a greater diversity of pollinators.

Examples of Flower Corolla Tube Lengths

Bee tongue lengths range from 2⁄25" (2 mm) to ½" (14 mm), depending on species.

• borage (*Borago officinalis*)	7⁄100" (1.9 mm)
• white clover (*Trifolium repens*)	1⁄10" (2.5 mm)
• lavender (*Lavandula angustifolia*)	¼" (6.0 mm)
• viper's bugloss (*Echium vulgare*)	¼" (6.7 mm)
• subalpine larkspur (*Delphinium barbeyi*)	½" (14 mm)

Example of a tubular flower: beardtongue (*Penstemon* species)

Generalists and Specialists

Generalist pollinators like honey bees and bumble bees can sip nectar from a wide array of flowering plants. And while most pollinators are generalists (approximately 75 percent of bee species), some have evolved as specialists and may depend on a particular plant species or a small group of plants. The specialist spring beauty bee (*Andrena erigeniae*) *only* collects pollen from two species of plants: Virginia spring beauty (*Claytonia virginica*) and Carolina spring beauty (*Claytonia caroliniana*). Without those two plants, the spring beauty bee may become extinct. Without the spring beauty bee, those plants may not be able to reproduce. Make sure to include plants for both generalist and specialist pollinators in your garden.

Butterfly caterpillars also have specialist relationships with plants, called larval host plants, discussed in chapter 3. A list of common host plants and the caterpillars they feed can be found on the author's website, www.ecobeneficial.com/PVG. Plants for specialist bees and beetles can be trickier to determine; not much has been written on this subject. Contact your local native nurseries and native plant society for help and refer to the recommended sources in this book.

Pollinator Sociability

When most people think of pollinators, they think of European honey bees, which are communal and live in large colonies, typically man-made beehives. Honey bees are social creatures and very different from the vast majority of pollinators; of the world's 20,000 bee species, it is estimated that 90 *percent* are solitary and nest by themselves or in small groups. This solitary behavior has implications for your Pollinator

Bumble bee queen nectaring on an early-blooming violet

Victory Garden because you need to create or facilitate areas where many individual bees can nest. Read more about providing bee habitat in chapter 2.

Seasonal Activity

Seasonal activity varies according to the pollinator species. Early-emerging bumble bee queens are active early in the growing season before many other insects are active. Bumble bees can tolerate cooler weather and somewhat rainy conditions in colder climates. When they emerge from their long winter's nap, they are hungry and looking for flowers. To help bumble bees and other early bees survive, include early-blooming plants in your garden, many of which are flowering trees and shrubs. Depending on where you garden, these early bloomers may include native willows, red maples, spicebush, serviceberries, redbud, witch hazel, trout lily, Dutchman's breeches, squirrel corn, wild geranium, spring beauty, and violets.

Other pollinator species appear later in the growing season; for instance, leafcutter bees emerge in warmer summer weather. Make sure they have enough to eat in your garden. In the fall, mated queen bumble bees need extra floral food before overwintering. Provide an uninterrupted succession of blooming plants throughout the entire growing season to provision a profusion of pollinator species.

Pollinator Life Span

Pollinator life spans can be quite different from one species to another. Most adult butterflies live for only a week or two, but an adult monarch or a mourning cloak butterfly may survive for nine months or more. Bee species have very different life spans, sometimes with significant differences within the same species. A honey bee worker might live just five to seven weeks, while a honey bee queen may live two to five years. Solitary wild bees may be active as adults for only three to four weeks, although their developmental stage is much longer at about eleven months. Simplify the provisioning process by planting for ongoing bloom, with a diversity of blooms at all times in the growing season, to feed pollinators with different life spans and seasons of activity.

Pollinator Foraging and Nutrition

Mining bee on golden Alexanders (*Zizia* species)

When pollinators forage (look for food from flowers), nectar and pollen are primarily what they seek. Bees are the main flower-visiting insects that drink nectar and also proactively collect and transport pollen. Pollinating flies and beetles may eat nectar, pollen, or flower parts, depending on the species. Butterflies and moths, with some exceptions, forage only for nectar, and they transfer pollen accidentally. Hummingbirds and bats visit flowers for nectar but may unintentionally digest grains of pollen.

A limited number of bee species collect other plant substances to line their nest cavities. Small resin bees gather plant resin that serves as a structural and waterproofing element with antimicrobial properties. Some other bee species forage for floral oils produced by plants with special floral glands. The bees mix these oils with pollen to line the brood cells in their nests, which also adds a protective, antimicrobial coating.

For the most part, nectar, pollen, and water are the essentials of a pollinator diet (with the exception of the larval stage of butterflies and moths).

Flower Diversity

Like humans, the majority of pollinators benefit from a varied diet. In most cases, the pollen and nectar they eat should come from multiple sources. Not all nectar and pollen have exactly the same composition or nutritional value. Think about providing a pollinator buffet by offering a wide selection of different flowering plants to your numerous pollinator guests. Don't let them go home hungry!

Floral Constancy and Flower Abundance

Flower diversity is important, but so is flower abundance. Some pollinators practice floral constancy, meaning they visit one plant species on a foraging trip. To forage successfully, these pollinators need a large enough quantity of a given flowering species. Research from the U.C. Berkeley Urban Bee Lab suggests that a 3 foot (0.28 m) square patch of the same plant species makes a good target for pollinators. Planting well for pollinators means striking a balance between plant diversity and plant sufficiency.

A pollinator buffet

Foraging Range

Not all pollinators can travel the same distance to find food. Tiny solitary bees may live within a few hundred feet of their nest; larger, quasi-social bees like bumble bees may venture a mile away, while honey bees can travel even farther. Butterflies and birds may travel even greater distances to get a meal. But travel has a significant cost for pollinators—they burn up calories and expend energy. Help pollinators save their energy by planting your garden intensively and convincing neighbors to create pollinator corridors that connect to your landscape and beyond.

Nectar

Most, but not all, flowering plants produce nectar. Some plants, including the majority of roses, produce little or no nectar but are important sources of pollen. Nectar is a sweet substance composed mostly of water and sugars, primarily glucose, fructose, and sucrose in various proportions. The amount of each type of sugar in nectar varies according to the plant and can vary within a species. Some pollinators seem to show a preference for a particular sugar type. Reportedly, hummingbirds and longer-tongued bees prefer sucrose, possibly due to its viscosity. Flies, beetles, and butterflies are thought to prefer lower levels of sucrose, while bats appear to be more attracted to nectar with glucose and fructose.

Nectar also contains amino acids and other components in small quantities. This sweet treat provides pollinators with an important source of energy that enables them to fly, breed, and stay warm. But the concentration of sugar in nectar can range dramatically, from 10 to 75 percent, which can affect a pollinator's ability to function. This is yet another reason to incorporate plant diversity in your garden so a variety of nectar resources are available.

Flowers secrete fresh nectar at varying rates, and this secretion and availability of nectar can occur over the course of a few hours or persist for a number of days. Some plants can produce nectar only during the day, while others can also produce nectar at night, making them especially valuable to nocturnal pollinators, like some nectaring moth species. Common milkweed (*Asclepias syriaca*) and evening primrose (*Oenothera* species) are two examples of plants that can deliver nectar at night, but choose plants appropriate to your region.

Temperature and weather conditions also affect nectar production by plants; very high or very low temperatures result in a decrease in nectar production. Excessive rain coupled with many cloudy days will cause nectar to become diluted. Drought stress significantly decreases the amount of nectar plants produce. Don't let plants wilt from lack of water, because pollinators will suffer the consequences. The best nectar delivery comes when plants have sufficient water as well as sufficient sun.

Pollen

Pollen is a source of protein for pollinators and provides other nutrients including fats, carbohydrates, minerals, and vitamins. Certain pollinators eat pollen, and some use it to feed their young. Pollen is essential for bee brood development (the development of young bees). Almost all bee species eat pollen, as well as some beetle and fly species, a few butterfly species, and some bird and mammal pollinators. Pollinators that proactively collect pollen for themselves or for their offspring expend a significant amount of energy to do so; be sure to have plenty of nectar plants in your garden to fuel them.

Not all plants provide the same quantity or quality of components in their pollen, nor is all pollen equally nutritious. Protein levels in pollen can range from 2.5 to 61 percent, yet a honey bee needs a diet with an average of 20 percent protein. If pollinators have access only to low-quality pollen, like that found in wind-pollinated plants, they will suffer; wind-pollinated plants will not get them to their minimum threshold.

The majority of pollen feeders seek pollen from a variety of sources, but there are also pollen specialists that have evolved with specific plants. Although most pollinators are pollen generalists, they may actually show a preference for particular plants.

Spicebush swallowtail on common milkweed (*Asclepias syriaca*), an excellent nectar source.

Bee covered in pollen grains

Bees need a clean source of water to drink.

Water

In addition to nectar and pollen, one resource rarely provided in a garden is water. Pollinators may use nectar as their primary source of hydration, but landscapes with a pond, stream, or river can be good water sources for pollinators, as long as there are shallow banks and the water is reasonably still. Even man-made water features, such as a tiny pond, can offer pollinators a drink, but design it to be pollinator friendly by keeping the same requirements in mind.

In the heat of summer, a honey bee colony can drink one quart (946 ml) of water or more each day. Gardeners can offer some simple water sources, which may be particularly important during droughts or very hot weather when nectar flow is low. Birdbaths are best kept for bird use, but shallow ceramic dishes, such as the saucers for plant pots, can be filled with small uncolored pebbles and topped off with water. This gives insect pollinators a safe place to drink without risk of drowning. Clever beekeepers have learned that automatic pet waterers can be used for bees too; just put small pebbles in the water dish to enable bees to drink safely from a dry perch.

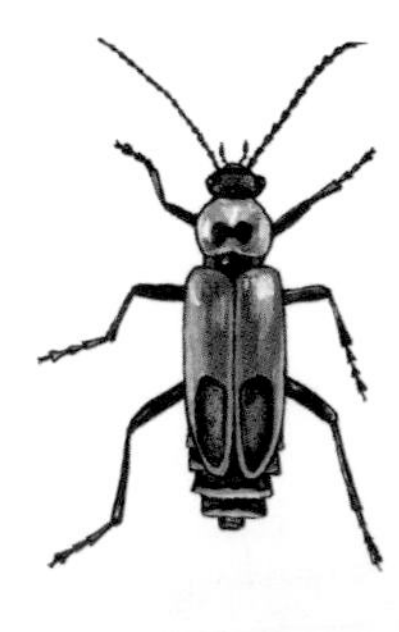

How Plants Attract Pollinators

Flowering plants have developed a number of traits to attract pollinators. Overall plant features such as flowering time, density of flowers, number of flowers, flower height, and spatial pattern of flowers can influence which pollinators will use a given plant.

An individual flower's color, shape, structure, size, fragrance, and availability of nectar and/or pollen also help determine what type of pollinator will visit. The combined groups of traits that predict the type of pollinator that will be attracted are known as pollination syndromes. Pollination syndromes are helpful guidelines for planting, but exceptions often occur. For more information, see pages 78 to 81.

Traits That Make Up Pollination Syndromes

- flower color
- flower shape
- flower fragrance/odor
- presence or absence of nectar guides
- nectar availability
- pollen availability

Nectar guides can be one factor in a pollination syndrome. They function as runways—patterns on flowers that direct pollinators, namely bees and butterflies, to the nectar. But not all pollinators are attracted by nectar guides; beetles, moths, flies, birds, and bats don't need them. Nectar guides are colored differently from the rest of the flower, visible to pollinators but often invisible to humans.

The complication for gardeners when choosing plants is that there is no "one size fits all" for pollinators. Bees favor violet, purple, blue, yellow, and white flowers, but hummingbirds favor red ones. Fragrance can play a tremendous role (or not), depending on the type of pollinator. For example, the intense perfume of a flower is lost on a bird because birds cannot smell, but a nocturnal moth may be enticed by a sweet nighttime bloom. Detailed information about the specific flower traits that attract pollinator types is given in chapter 4.

Nectar guides are visual cues to some pollinator species (spring beauty, *Claytonia* species).

2 Providing Pollinators with a Place to Live

Habitat is a term that frequently appears when pollinators are being discussed. But what *is* habitat? Simply put, habitat is the natural environment where an organism, in this case a pollinator, lives, and where it can find the resources it needs: food, shelter, resting places, and safe places to mate and reproduce. Habitat includes more than flowers.

Pollinator Victory Gardens provide multiple types of plants and pollinator habitats.

Understanding Habitat as an Ecosystem

Any residential, commercial, or institutional landscape is more than a combination of trees, shrubs, perennials, and (likely) lawn; it's an ecosystem where every living thing is interconnected. And *your* ecosystem can provide habitat for many types of wildlife, including pollinators.

Most of us have landscapes that can function, at least to a limited degree, as pollinator habitat. But there is a tremendous difference between quality pollinator habitat and poor pollinator habitat. A landscape dominated by a large turfgrass lawn will never be a high-quality pollinator habitat; you could improve it by allowing flowering weeds to exist in the lawn, but that still won't be enough. By considering what pollinators need, you can create habitats that are exceptional versus merely marginal.

Key Elements of Pollinator Habitat

- forage and floral resources (nectar, pollen, floral oils and resins)
- nonfloral plant resources (larval host plants, leaves for nesting materials, propolis)
- clean source of fresh water
- areas for egg laying and nesting
- areas for sheltering and overwintering
- areas for resting and warming

What Pollinators Need in a Habitat

Habitats don't exist in a vacuum; think of your entire landscape as habitat, not just the perennial garden. Yes, flower-filled perennial areas will see a lot more obvious pollinator activity, but pollinators use many other parts of your landscape that you may not notice. Trees and shrubs can offer flower rewards, food for butterfly and moth caterpillars, nesting spots for pollinating birds, resting and overwintering sites for insects, and protection from strong winds. Areas of bare or lightly vegetated soil, brush piles, and plants with hollow stems provide places for some pollinators to nest. Bunches of native grass, leaf litter from trees, and piles of twigs can serve as overwintering sites for some pollinator species.

When you improve your ecosystem for pollinators, the overall ecological health of your landscape improves, benefiting all plant and animal life. To keep pollinators and other wildlife healthy, your landscape, ideally, should be pesticide free.

Habitat Regionality

Pollinator habitats in Arizona should not look like pollinator habitats in New Jersey. Habitats are most successful when they reflect the country and the region in which they exist. Climate, geography, indigenous pollinators, and the native plants they have evolved with, differ from region to region. Using resources in this book's appendix, you can determine the ecological region you live in, and the plants and animals that are native to that region. This provides some context for the type of pollinator habitats most suitable for your locale. Remember, evolutionary connections matter.

Native willows (*Salix* species) are larval host plants for over 450 species of caterpillars and an early source of nectar for bees.

Creating Habitat for Egg Laying and Nesting

All pollinators need a safe place to reproduce, which is one of the most important aspects of your Pollinator Victory Garden. Simply because you see pollinators nectaring in your landscape does not mean that the conditions are right for egg laying and nesting.

Butterfly and Moth Habitats for Egg Laying

Butterflies and moths don't make nests like some other pollinators, but they do need a place to lay their eggs. The majority of butterflies and moths lay eggs on specific plants that have evolved to feed the caterpillars that emerge from these eggs. These plants are called larval host plants (more about this in chapter 3). If caterpillars don't have their larval host plants, they cannot survive.

What makes good butterfly and moth egg-laying habitat? Larval host plants that are pesticide free. If you plant only flowers for adult butterflies and moths to nectar on and don't have larval host plants, you don't have a pollinator garden to support butterflies and moths.

Native Bee Habitats for Egg Laying and Nesting

All native bees in North America build nests, except cuckoo bees, which lay their eggs in the nests of other bee species. Since the vast majority of native bees are solitary, most bees nest alone, but bumble bees are an exception; they nest in small groups of fifty to four hundred. Female native bees build nests and provision the nests with food for their young, while male bees find other spots to rest and sleep on, including flowers, stems, and twigs.

The majority of native bees nest in the ground.

Native bees nest either in the ground or in cavities. The majority (70 percent of all native bee species in North America, or 2,800 of the 4,000 species of native bees) are ground nesters. The other 30 percent, including bumble bees, are cavity nesters. There are a lot of bees to provide habitat for in your pollinator garden! You may have hundreds or even thousands of individual bees using your landscape. The good news is that you don't have to do much to give them a place to live.

Butterfly Bush—Not

Unfortunately, many gardeners assume that butterfly bush (*Buddleia davidii*), and perhaps a few other flowering plants, are all they need for butterflies. Butterfly bush is a plant that evolved in Asia; it does not serve as a larval host plant for *any* North American butterfly species. It does provide nectar, which many butterflies enjoy, but pity the poor female butterfly nectaring away on butterfly bush when it's time to lay her eggs (known as ovipositing). Instinctively she realizes that it's not a plant that will feed her young, and she won't lay eggs there. If there aren't any larval host plants nearby, all that potential progeny is doomed.

In North America, butterfly bush falls short as a true butterfly plant. "Nectar bush" more accurately describes what the plant actually does—it delivers nectar, period. There are many native plants that can do that too. Another reason to think twice about butterfly bush: It can be a prolific re-seeder and is now listed as invasive plant or a species of concern in several states.

Do a little research on the butterflies that occur in your region and their preferred larval host plants, and then plant those. A list of recommended books can be found in the appendix. A list of common butterflies and moths and their larval host plants can be found at www.ecobeneficial.com/PVG.

Examples of Ground-Nesting Bees and Cavity-Nesting Bees

GROUND NESTERS	CAVITY NESTERS
• mining (digger) bees	• bumble bees (above or below ground)
• sweat bees	• mason bees
• cellophane bees	• leaf-cutter bees
	• yellow-masked bees
	• carpenter bees
	• resin bees

Providing habitat for ground-nesting bees

The majority of native bees are ground nesters that need easy access to soil to excavate their nests. Reserve existing patches of bare soil in a sunny spot or create an expanse of bare soil—even a few square feet (m) can work. Sandy or loamy soils are top choices; rich soils and clay soils are more difficult for bees to excavate and may get waterlogged. A sunny, quiet location, free of mowers, noisy equipment, and foot traffic, is ideal. If naked soil seems too unsightly, you can plant the area very sparsely and bees may still use it. Best to leave the area bare but screen it with some flowering shrubs. Just make sure not to shade out the bee habitat.

Avoid using mulch or landscape fabric in bee nesting areas as both will repel bees. All habitat should be left quiet and undisturbed, with no tilling or cleaning up, even in the off season. Think of this area as wild habitat instead of a manicured garden. If you succeed, you may support hundreds of bees underground.

Providing habitat for cavity-nesting bees

Thirty percent of native bees are cavity nesters that use an array of resources for their nests. Many like pre-existing cavities such as abandoned beetle tunnels in fallen logs, dead trees, and stumps; others may use gaps in stone walls, abandoned bird nests, hollow trees, spaces within rotting wood, dense brush piles, or plant debris. Some bees even nest inside pithy (spongy) plant stems.

Old beetle tunnels in a dead tree offer habitat for cavity-nesting bees.

Ceratina bees nesting in a pithy plant stem.

Tree snags

Consider preserving dead or dying trees as bee habitat. Such trees are referred to as snags and can provide valuable habitat for pollinators and many other creatures. A hollow expanse in a dead tree may give bumble bees or even feral honey bees a safe home. Beetles like to tunnel in dead wood, leaving a future home for cavity-nesting bees. Any snag located well away from your house, where it won't harm anyone or anything if it falls, is worthy of keeping. If the snag is too tall for your comfort, cut it back to a height that is manageable; wildlife will still use it. Leave fallen logs in place as a habitat resource. As they decay, they often become quite beautiful with an abundance of plants growing in the rich medium. Just consider them nature's free planters.

Pithy stems

Pithy stems from plants such as elderberry, raspberry, hydrangea, and Joe Pye weed are highly useful to some cavity-nesting bees. You can leave the dead stems from plants like these in place over the winter or collect and bundle them for bee habitats. As a general practice, leave any perennials standing in the garden through winter to provide habitat for insects and seed for birds. They can then be cut back in early spring. It's best to stagger your early season stem cutting over a period of time, at least a week or two, to ensure that temperatures have warmed sufficiently to allow insect visitors to become active.

Brush piles can be good habitat for bumble bees.

Brush piles

Dense brush piles are a good potential habitat for bumble bees. You may not have any because conventional wisdom dictates that we clean up our landscapes, removing any dead trees and branches. This practice, however, is counterproductive to good ecological practices. If you don't have brush piles in your landscape, create them! Brush piles don't have to be huge; whenever you prune woody plants to remove dead wood, keep the branches. You can secure them tightly in place or tie them together. To successfully shelter bees, the piles must be fairly tight and dense, providing cavities that shelter bees from rain. If you are handy, try making a wattle fence, which is a wooden fence made by weaving thin branches together to form a lattice, which will also provide shelter.

Man-made bee hotels can serve as habitat for cavity-nesting bees.

Man-made habitats for cavity nesters

Man-made habitats for cavity nesters can be made or purchased. Wood blocks drilled with holes in a range of diameters are called nest boxes. Hollow plant stems can be bundled together to form stem bundles. There is a bit of an art and science to these artificial habitats, including maintaining proper hygeine, so check the appendix for suggested resources.

Providing Habitat for European Honey Bees

European honey bees (*Apis millifera*) have an arrangement for nesting and raising their young that's very different from that of wild native bees of North America. When they're managed by a beekeeper for honey production or pollination services, honey bees live in man-made bee-hives that usually look like stacked boxes. They lay their eggs and raise their young in those hives.

There are also feral honey bees that are no longer managed by beekeepers. These feral honey bees often live and reproduce in tree cavities as a large group, which is yet another reason to leave dead or dying trees in place as long as the trees pose no hazard.

Honey bees are communal, the most social bees, and may live in colonies of 50,000 or more individuals depending on the season. You don't have to be a beekeeper to help honey bees. By keeping your landscape pesticide free and filled with flowering plants, you can support both native bees and European honey bees. Since there is competition for resources among native bees and honey bees, plant intensely if you keep honey bees or have neighbors who do.

Vegetative windbreak with a nectar bonus—coral honeysuckle (*Lonicera sempervirens*)

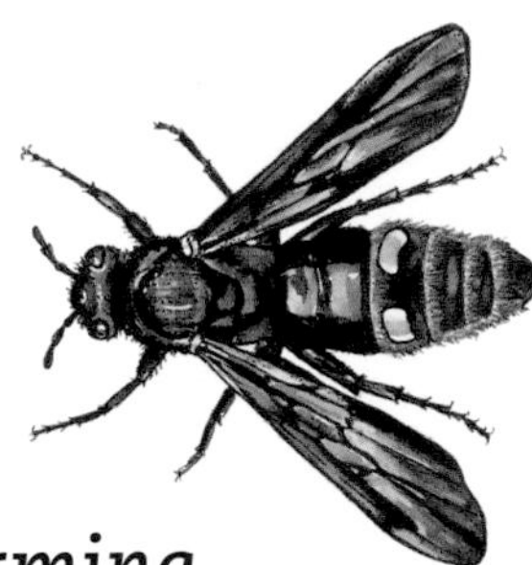

Providing Habitat for Sheltering and Warming

In their active seasons, pollinators require safe places to rest, warm their bodies, and shelter in bad weather. When the weather is fine, insects like bees, moths, and wasps may rest or sleep on flowers overnight. In the case of bees, they hold on to the flowers with their legs or their mandibles (jaws) and rest. Sometimes bees even sleep in flowers that close up for the night, which may be a way to stay safe from predators. We can design landscapes to help them with all of these needs.

Butterflies don't have nests, and they don't even sleep—they rest with their eyes open. Butterflies look for a protected place to rest among foliage, or they hang upside down from leaves or twigs of trees and shrubs. While butterflies are active during the day, most moths are nocturnal and find similar protected places to shelter during the day.

Sheltering and Windbreaks

When weather is bad, pollinators are quite vulnerable. Consider the small size of most pollinators and how exposed they are to high winds. A monarch butterfly can weigh as little as 0.009 ounce (255 mg), while a bee might weigh only 0.003 ounce (85 mg). Pollinators that have nests, like bees, wasps, and hummingbirds, may not be able to return to their safe haven in time to avoid winds. Butterflies are especially vulnerable to wind because they lack nests. But with some planning, you can plant vegetative windbreaks to help shelter and protect pollinators on windy days.

Windbreaks serve many purposes. When protected from wind, pollinators not only have a greater chance of survival but also are able to conserve energy, which allows them to forage more on flowers. Windbreaks also provide conditions that are a bit warmer, which allows cold-blooded insect pollinators like bees to be active for longer periods.

Native grasses can serve as windbreaks and bee nesting sites.

Plants for windbreaks

Trees and shrubs are particularly effective as windbreaks. Flowering woody plants have the extra benefit of providing additional nectar and pollen, further expanding the forage habitat. In especially windy areas, the cover of evergreen trees and shrubs can be useful, and some also flower.

Tall native grasses can serve not only as windbreaks but also as bee habitat. Native bunchgrasses, such as prairie dropseed, are good overwintering sites for bumble bee queens, which may nest at the base of the grass. Many native grasses are larval host plants for skippers, a large butterfly family. Native grasses perform even more ecological functions as their deep roots hold moisture and reduce flooding and soil erosion.

Creating a windbreak

There is no magic formula or exact design for creating vegetative windbreaks for pollinators; it can be accomplished with simple plantings. Windbreaks near pollinator nesting sites are helpful, but make sure not to site plants where they cast shade on nesting habitat. Warm-season native grasses can accomplish this as they are not too tall but still tall enough to provide shelter.

Assess the windy areas of your landscape, especially around foraging habitat. Staggered rows of diverse flowering trees and shrubs can be useful in these locations. The more diverse the planting, the better, to avoid monocultures of shrubs positioned side by side like little soldiers. You are looking to create a windbreak, not a green wall! Monocultures limit ecosystem functioning (your landscape is an ecosystem) and a monoculture planting may be wiped out by a single pest or disease.

Butterflies, like this red admiral, bask on rocks to warm their bodies.

Habitat for Warming

Pollinating insects are cold blooded and need to warm up to become active. Although some bees, like bumble bees, can tolerate less-than-ideal weather conditions, honey bees may only be active at temperatures of 54°F (12.2°C) or higher, and butterflies reportedly need a temperature of 55°F (12.7°C) or higher to fly. Providing habitat in sunny spots helps them warm up on days when temperatures are questionable; this is a major reason for planting forage in sun-filled areas.

Rocks can be another aid for warming and are underused as a design feature in gardens. Boulders and large rocks, which stay warm longer, can be nicely positioned to complement plantings and create more visual garden interest. Place them in sunny locations to create a basking area that many pollinators will appreciate.

Loose bark of shagbark hickory (*Carya ovata*) offers habitat to pollinators like butterflies.

Creating Overwintering Habitats

In colder climates, most pollinating insects die before winter sets in. But many do survive over the winter using a variety of survival strategies. Here are some ways various types of pollinators overwinter.

Butterflies and Moths

Most butterflies overwinter as caterpillars (larval stage) or as a chrysalis (pupal stage), but some species overwinter as eggs or adults. Many species undergo diapause, a period of suspended development that usually occurs during cold winter months. Mourning cloaks, anglewings, tortoiseshells, and eastern comma butterflies can overwinter as adults in colder climates as they have special chemicals in their bodies akin to antifreeze that prevent them from freezing. Some butterfly species migrate, including monarchs, which migrate the longest distances; shorter-distance migrators include painted ladies, buckeyes, and cloudless sulphurs.

Leaf litter can be valuable overwintering habitat for some pollinators.

Adult overwintering

Butterflies and moths that overwinter as adults often shelter in the crevices of dead trees, in hollow logs, between large gaps in rocks, or sometimes in leaf litter. Some adult butterflies find cover under the exfoliating bark of trees. Shagbark hickory, a tree native to the eastern United States, has loose bark that shelters overwintering butterfly adults. Piles of logs can also make excellent winter habitat for adult butterflies.

Caterpillar overwintering

Many caterpillars and almost all moth caterpillars overwinter in leaf litter; pupae typically overwinter suspended from twigs or branches but may also be found in piles of dead leaves. Leaf litter is a refuge for many insects, not only butterflies and moths, and that's a great reason to leave fallen leaves in place in your yard. Resist the urge to chop up or mulch those leaves, as you may unwittingly kill adult or immature butterflies or moths. Raking and mowing in fall may also remove leaf litter winter habitat that is so beneficial to these creatures.

Honey bee on wild plum (*Prunus americana*)

Honey Bees and Bumble Bees

Honey bees are the most social bees and overwinter in the same location where they nest during their active season, which is either a man-made hive or, in the case of feral honey bees, a hollow tree or similar cavity. If weather turns warm in winter, honey bees emerge, looking for food.

Bumble bees form much smaller colonies during their active season than honey bees but become solitary in winter. In the fall, a bumble bee colony produces new queens and male bees. Once queens and males mate, the newly mated queens survive but the other bees in the colony die. The mated queen finds a new nest, often a small underground spot obscured by leaf litter, where she overwinters by herself. Sometimes she nests under rocks or rotting logs or at the base of native grasses. Protect her overwintering habitat by leaving leaves and logs in place. Overwintering bumble bee queens, like overwintering adult butterflies, produce a natural chemical called glycerol that prevents their bodies from freezing in cold weather. In early spring the queen emerges from hibernation and finds a nesting site to lay her eggs.

Solitary Bees

Solitary bees live for about a year, most of which is spent in their developmental stages of egg, larva, and pupa. These bees are active as adults for only a short time—perhaps three to four weeks. Unlike honey bees and bumble bees, solitary bees of most species die as adults before winter arrives. Solitary female bees seal their eggs inside a burrow or nest cavity provisioned with a supply of pollen and nectar. Preserve their habitat undisturbed so new bees can successfully emerge in spring or summer.

Surviving solitary cavity-nesting bees, including mason bees, leafcutter bees, and small carpenter bees, overwinter as adults in their original nests in a state of diapause, or suspended development. Mason bees are particularly interesting since they may overwinter in waterproof cocoons. To help cavity-nesting bees, grow plants with pithy, spongy stems, such as elderberry and Joe Pye weed; keep trees and shrubs that have hollow voids or beetle tunnels, and consider providing bundled hollow tubes or man-made bee hotels (see the appendix for resources).

A Pollinator Victory Garden is full of flowers, but it's more than just a place for pollinators to eat. A flower garden devoid of good nesting, sheltering, and wintering sites is not a Pollinator Victory Garden; it's merely a floral buffet. Give pollinators a safe, pesticide-free home where they can live and reproduce.

3 Providing Pollinators with Food to Eat

Habitat consists of a place for pollinators to live as well as a place for them to eat and get the nutrition they need to flourish. These habitats may be the same, can be overlapping, or may be more disconnected as is the case for honey bees, which will fly several miles, if necessary, to obtain food. In managed landscapes, the place to live and the place to eat (both habitats) should overlap and are equally important. In most cases, you can provide pollinators with both a place to live and a place to eat in the same general area.

Pollinator habitat with plenty of pollinator forage

What Pollinators Eat

Forage is the word for the food supply from flowers that pollinators eat. Foraging habitat is the dining room of your landscape, the parts of the garden where pollinators can find a meal.

Nectar, Pollen, and Other Food

The main pollinator foods are nectar (for energy) and pollen (for protein), which are produced by most flowering plants. Some flowering plants, such as St. John's wort and shooting star, produce only pollen, but that still has value for pollinators. Roses, which are often mentioned as pollinator-friendly plants, have little nectar but reward pollinators mostly with pollen. For pollinators, stick with native roses that have simple floral structures and abundant pollen. Avoid hybrid roses, which often have little or no nectar or pollen, as well as more complex petal structures that make any forage hard to access.

Common Eastern bumble bee foraging on common sneezeweed (*Helenium autumnale*)

In addition to nectar and pollen, some pollinators seek out nonfood resources from flowers. Floral oils produced by a limited number of flowering plants have evolved with specialist bees, which use them in their nest cavities. Plant resins from flowers or plant stems are collected by some bee species for the same purpose.

Differences in What Pollinators Eat

While honey bees and wild native bees eat both nectar and pollen, not all pollinators do. Most adult butterflies eat only nectar, while some neotropical butterflies feed on both nectar and pollen. Other species of butterflies prefer nonfloral foods, including tree sap. The majority of adult moths eat nectar, with the exception of species like wild silk moths that eat nothing at all, gaining the nutrition they need in their larval stage.

Some beetle species specialize in eating pollen while others may eat flower parts, like petals, as part of their diet. Wasps are primarily carnivores, mostly consuming insects, but they may also eat nectar, especially when food is scarce. Hummingbirds and pollinating bats feed on insects, nectar, and sometimes pollen.

Native plants such as coneflower (*Echinacea purpurea*) often attract a diversity of pollinators.

How Pollinators Feed

Most pollinators feed multiple times a day and, in the case of bees, may make dozens of foraging missions daily. Some pollinator species are specialists on particular plants, while others are generalists, feeding on many different plants. An interesting exception is the bumble bee; although bumble bees are usually generalist foragers, individual bees in a bumble bee colony may specialize on certain plants.

Some pollinators, including bumble bees, honey bees, and butterflies, have a feeding behavior called floral constancy, also known as floral fidelity. On each foraging trip, they look for one species of plant from which to eat. This is a clue to the gardener to plant more than one of any specific plant; otherwise, pollinators practicing floral constancy will not get enough nutrition and will expend more energy on more foraging missions. Pollinators benefit from groupings of plants at least 3 feet (0.28 m) square. Not all gardens have room to accomplish this—do the best that you can.

Scorpionweed (*Phacelia* species) is particularly attractive to honey bees.

Columbine (*Aquilegia candensis*) is favored by hummingbirds.

Choosing Plants for Pollinators

Choosing the best plants for your Pollinator Victory Garden is easier when you have an understanding of plant categories and the reasons for choosing particular plants over others. Native plants, nonnative plants, naturalized plants, straight species plants, cultivars, and hybrids all have different implications in a pollinator-friendly landscape.

"Plant the right plant in the right place" is a gardening mantra to remember. The plants you choose should be appropriate for your region and suitable to the particular site you are planting. And the plants you select should appeal to pollinators, at least some pollinators. There really is no "one size fits all" plant, as not every pollinator is attracted to or can use every pollinator plant.

Once you learn the key points of pollinator plant selection addressed in this chapter, you can make the best choices for your Pollinator Victory Garden. Spoiler alert: For best pollinator results, favor a diverse array of straight species plants native to your region and suitable for your landscape conditions.

Research suggests that native plants are four times more attractive to pollinators than nonnative plants. The evolutionary connections between native species are profound in some cases, not to mention the fact that native plants can offer more to an ecosystem than just nectar and pollen, and are best adapted to regional conditions.

Using native perennials and woody plants

The benefits of native plants are undeniable in a pollinator garden, but they are grossly underutilized in most landscapes. Your choices of natives include not only perennials but also native trees and shrubs. Think of your garden as your entire landscape, not just a discrete planting area where flowering perennials live. Masses of flowering native perennials will be a boon to pollinators but so will the native trees and shrubs you plant.

Many native trees and shrubs are larval host plants for butterflies and moths. The more host plants you can include in your pollinator landscape, the greater the number of butterfly and moth species you will support. Flowering native woody plants offer an abundance of flowers with a volume of bloom that can be hard to replicate with perennials, particularly in a small yard.

Incorporating more natives

Consider the areas of your landscape where you may have underperforming shrubs that could be easily replaced by native pollinator-friendly substitutes. A nonnative shrub with blooms that no pollinator visits is a good candidate to replace with a pollinator-enticing native alternative—perhaps a ninebark, a ceanothus, or a summersweet, depending on your region. Many lifeless foundation plantings are waiting to be transformed into pollinator buffets. Trees and shrubs that are not healthy can also be removed to make room for robust and ecological native pollinator plants.

Oaks (*Quercus* species) serve as larval host plants to the greatest number of caterpillar species.

When selecting native plants, consider the layers of the landscape and the native species you see in natural areas around you. Try to reflect those layers and species in some of your landscape. In much of the northeastern United States, natural areas have layers that start with canopy trees, transition downward to subcanopy trees, tall shrubs, and smaller shrubs, and then descend to the herbaceous layers of flowering perennials, grasses, sedges, and ground covers. Even prairies and deserts have layers, although they are harder to see. Emulate the natural layers you see in your region in your own landscape.

Bee nectaring on summersweet (*Clethra alnifolia*), a native plant

Nonnative Plants

A nonnative plant is one that has been introduced with human help, either intentionally or accidentally, to a new place or new type of habitat where it was not previously found. Another term for nonnative plant is *alien plant*; exotic plants are nonnatives that come from a different continent. Not all nonnative plants are invasive, nor do all nonnative plants naturalize; in fact, some nonnative plants are well behaved. Although flowering nonnative plants may provide nectar and pollen to generalist pollinators, they may not function ecologically in other ways that native plants do—as larval host plants, as preferred nesting sites, as sources of appropriate nutrition for migrating birds, and so on.

When considering nonnative plants with which you are not familiar, first check your state's list of invasive species to determine if a plant has been declared invasive or a species of concern, or if it is on a watch list. Avoid planting *all* such plants and look for native alternatives. Many invasive plants were commonly used in gardens and municipal plantings before their detrimental effects were understood. (And some are still being sold in the trade, so beware.)

Invasive plants, such as kudzu, choke out native plants and eliminate native habitat.

Naturalized Plants

A naturalized plant is a nonnative plant that can reproduce and maintain itself over time without human help. Even with time, though, naturalized plants do not become native members of the local plant community. Invasive plants are a subcategory of naturalized plants, and those are the ones you really need to eliminate.

Naturalized plants are best avoided in your landscape. Some naturalized plants may become problematic in the future if they disrupt ecosystems, displacing native plants that would otherwise be present. If you want plants that spread, choose native plants that have this trait.

Invasive Plants

An invasive plant is a plant that is nonnative, is able to establish on many sites, grows quickly, and spreads to the point of disrupting ecosystems. However, not every plant with an aggressive growth habit is an invasive plant, as even some native plants can be aggressive, and this can be useful in certain garden situations.

The U.S. Government defines invasive species as species nonnative to the ecosystem under consideration and whose introduction causes, or is likely to cause, economic harm, environmental harm, or harm to human health. Plants that are invasive in one part of the United States may not be problematic, at least for the time being, in another part of the United States. Common invasive plants in the Northeast include Japanese barberry, oriental bittersweet, Norway maple, and Japanese honeysuckle. Common invasive plants in California include black acacia, giant reed, Mexican feathergrass, and tree of heaven.

Many plants that have become invasive over time were actually introduced through horticulture long before there was any suspicion that these plants could cause problems outside of their native range. Most states in the United States have their own invasive species list. Do not plant those plants, and remove any you may already have. Also avoid plants listed as species of concern but that are not yet defined as invasive.

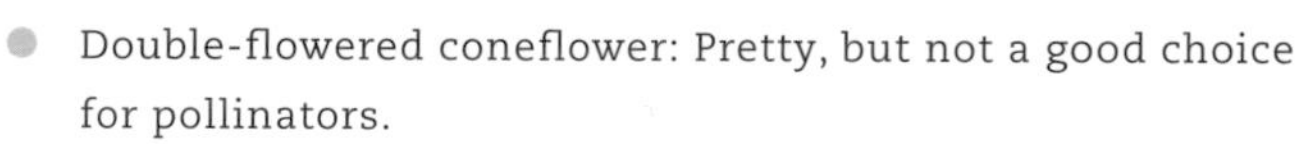

Double-flowered coneflower: Pretty, but not a good choice for pollinators.

Native coneflower (*Echinacea purpurea*): A great choice for pollinators.

Plants to Leave Out

When considering plants for your Pollinator Victory Garden, start with the easy decisions about plants that should be left out. You will likely see many of these plants in other gardens—but don't you make the same mistake. Following are the three groups of plants you can eliminate or minimize for pollinators.

Eliminate These Plants for Pollinators

1 **Eliminate plants listed as invasive or species of concern on your state's lists.** Invasive nonnative plants are so robust they can cause environmental or economic harm. These are the plants you often see choking out native plants, especially in disturbed areas, sometimes creating great monocultures, happily free of the pests and diseases that keep them in check in their country or region of origin. Invasive plants cause real ecological harm by displacing the native plants that native wildlife has evolved with and depends upon.

Unfortunately, some invasive nonnative plants are quite compelling to pollinators. Pollinators are opportunists and will use the floral resources (nectar and pollen) that are available in the absence of native or well-behaved nonnative plants. Keep the overall health of your ecosystem in mind and keep invasive nonnative plants out of your plant palette.

2 **Eliminate plants with double flowers or those that have lost most of their nectar and pollen from overbreeding.** Plants that sport double flowers typically have little or no nectar or pollen for pollinators. The extra petals replace the plant's reproductive parts, minimizing or eliminating the floral resources pollinators need, which are nectar for energy and pollen for protein. Double flowers may be visually captivating to humans, but they are useless, or close to it, for pollinators.

Often a plant that bears little resemblance to its naturally occurring counterpart isn't a worthy pollinator plant. If you're in doubt, ask at a garden center or nursery whether the plant you are considering is sterile. If you are fond of an unusual plant that is a sterile hybrid, you can include it elsewhere in your landscape, but consider it an ornamental garden object, not a pollinator plant or a plant with significant ecological value.

3 **Minimize wind-pollinated plants.** While some wind-pollinated plants, such as hazelnuts, may provide limited low-quality pollen to pollinators, wind-pollinated trees and shrubs are best reserved as larval host plants for butterflies and moths and as nesting and food sources for birds and mammals. A garden of mostly wind-pollinated plants is a pollinator desert. Trees and shrubs that are wind pollinated can provide important resources but don't provide enough food for pollinators.

Some wind-pollinated perennials *are* worthy of planting in a pollinator garden. Warm-season native grasses are an example. Although they do not provide nectar or pollen, native grasses are often larval host plants for skippers (a type of butterfly) and have the additional value of providing a good windbreak for pollinators, as well as being a potential nesting site for bumble bees.

Fritillary butterfly nectaring on butterfly weed (*Asclepias tuberosa*)

More on Choosing Plants: Straight Species, Cultivars, and Hybrids

When you go plant shopping, labels and plant names can be confusing—and you may not always be sure of what you are buying. If you know the differences in how plants have been propagated or bred you can make the best choices for pollinators.

Behind every successful Pollinator Victory Garden are plants that have the characteristics that attract pollinators and deliver the resources they need. A flowering plant that has little or no nectar or pollen may be beautiful, but it has little value in a pollinator garden. A plant that is cultivated for its peculiar flower color or other significant deviation from the species may not be recognized by the pollinators you are trying to attract. Mother Nature really does know best—follow her lead when choosing plants.

Botanical Names

With regard to plant names, the only way to truly identify a plant is to use its botanical name. Those of us who suffered through Latin classes in school can finally put some of that education to good use. Common names for plants are not always helpful, as they can vary significantly from country to country, state to state, and region to region. Latin is your friend when it comes to plants because it ensures that you get exactly the plants you want.

Botanical names consist of two words in Latin. The first word, always capitalized and italicized, is the genus; the second word, always lowercase and italicized, is the specific epithet. The two words together form a plant's name, also known as the plant species. *Asclepias tuberosa* is an example of a plant species. This particular plant has myriad common names, including butterfly weed, butterfly milkweed, pleurisy root, orange milkweed, chigger flower, fluxroot, and orange root. Botanical Latin makes it clear, no matter where you live, what a given plant really is.

As plant DNA analysis becomes more sophisticated, botanical names sometimes change. Plant taxonomists (the folks who identify, describe, classify, and name plants) may rename a plant to better align with its true lineage and relationship to other plants. This makes complete sense—until you go to a garden center looking for *Aster laevis* (smooth aster) and can only find *Symphyotrichum laeve*, the new botanical name for the same plant. But, growers, nurseries, and garden centers are not always consistent about using new names on plant tags. Go plant shopping armed with both old and new botanical names!

Plant Choices

Genetic diversity of plants within a species is also important for ecological health, but not all available plant choices offer this diversity. As with most things, there are trade-offs when making plant choices. Go to the garden center armed with the information you need!

This means understanding the differences between straight species plants, cultivars, and hybrids. From an ecological perspective, straight species plants offer the greatest biodiversity and the biggest ecological impact, both important factors for healing our challenged environment. Here is a look at the differences between the three types.

Phlox paniculata 'Jeana', a nativar of garden phlox that is a clone. It does attract many butterflies and moths.

Straight Species Plants

A straight species plant is one that occurs naturally without breeding or hybridization. "Open-pollinated" plants grown from seed have the greatest genetic diversity. Give a preference to regional, straight species native plants that are open pollinated (meaning, naturally pollinated by insects, wind, or water) for their boost to biodiversity and their adaptation to the local region.

An example of a straight species plant is *Penstemon hirsutus* (hairy penstemon). Some plants may have a naturally occurring subspecies or variation of the plant, such as *Penstemon hirsutus* var. *pygmaeus*, a dwarf form of this species.

You may also see a subspecies of a plant designated as "ssp."

Cultivars and Nativars

A cultivar is a cultivated variety of a plant, often selected for a desired trait, such as pest or disease resistance, flower color, fruit color, foliage color, habit, size, and so on. A cultivar may be propagated by seed or by vegetative cutting (resulting in a genetic clone). Cultivars are designated by single quotes—for example, *Echinacea purpurea* 'Magnus'. A cultivar name is not italicized.

A nativar is simply a cultivar of a native plant. Unfortunately, this is often the only type of "native" plant available at nurseries. Nativars are somewhat controversial since they have less genetic diversity and potentially have some loss of ecological function compared to straight species plants.

Many nativars are genetic clones that have been propagated from cuttings, while others are open-pollinated and grown from seed. Clones are just what you imagine—think Dolly the sheep—and they lack genetic diversity. Plant labels typically do not disclose whether a plant has been grown from seed or from a cutting. Do a little digging online when buying plants.

In terms of an ecological best bang for the buck with the most biodiversity, straight species plants are first choice, open-pollinated nativars are next best, and cloned nativars come in third.

Hybrids

Hybrid is a word that often refers to the intentional breeding of two different plant species to create a new plant. Hybrids are designated by the absence of the second Latin word in a plant's name (its specific epithet) or the presence of an "x" symbol. For example, *Baptisia* 'Purple Smoke' is a hybrid of *Baptisia alba* (white wild indigo) and *Baptisia australis* (blue false indigo). Hybrids are controversial in the world of native gardening due to their loss of genetic diversity, although some natural hybrids do occur without human intervention. Some hybrids, but not all, are also sterile. In a Pollinator Victory Garden, hybrids are not the best choice.

Plant Selection Tools and Tips

Proper plant selection is key for any garden including a Pollinator Victory Garden. For best plant health, pollinator support, and positive ecological impact, emphasize plants that are native to your local area and appropriate for your landscape conditions.

Stay Local

Plants form the basis for local ecosystems and have evolutionary connections to an area's wildlife, including pollinators. Yes, generalist native pollinators may nectar on nonnative plants and nonnative honey bees may sip nectar from native plants, but ecosystems are complex and the depth and extent of evolutionary connections and ecological functions may not always be obvious. Native nectar plants that also function as larval host plants are an example of these complex evolutionary connections.

For best success, plant native plants that native pollinators have evolved with in your area. Look for locally grown native plants or seeds when shopping. As a bonus, native plants will boost the overall ecological health of your landscape. Don't worry if you also want to support honey bees; many native plants will feed these generalist pollinators too.

Match Plants to Landscape Conditions

The old saying "Plant the right plant in the right place" was likely coined by a gardener who learned the hard way. This pithy phrase is often followed by the consoling adage "You don't know a plant until you've killed it." Better to plant the right plant in the right place to begin with! No matter how fantastic a pollinator plant is, if it is not appropriate for your landscape

Chokecherry (*Prunus virginiana*), a native plant that is a larval host and a good nectar plant.

conditions, *don't plant it*. It's just a matter of time before a plant that requires full sun and fast-draining soil will fail when it's planted in a shady garden with soggy clay soil. Research any plant you are considering for your pollinator garden to determine if it is suitable for your own landscape conditions. There are many good resources noted in the appendix of this book to help you research plants.

Goldenrod (*Solidago* species) is an important source of late-season nectar.

Understand Plant Regionality

Regional plant lists can be a helpful starting point when selecting plants for your Pollinator Victory Garden. But the way regions are defined can vary based on the source. As a start, see if your state is included in the region being discussed.

Don't be surprised if you see the same plant genus (collective name for a group of related plants) occurring in multiple regions of North America. For example, the genus *Solidago* (goldenrod) is found in every region of the United States, accounting for more than 100 species. But not every species of goldenrod occurs in every region or in the same geographic conditions. Some species have very large ranges and are adapted to a variety of ecological regions (ecoregions).

Solidago altissima (tall goldenrod) has a wide range and is found in most regions of the United States. Other goldenrod species have extremely limited ranges—for example, *Solidago albopilosa* (whitehair goldenrod) is found only in Kentucky.

Because the geography, ecology, and growing conditions can be very different within a particular state or region, not every plant listed for that state or region is suitable for every landscape. Gardening in California underscores this point. California's ecoregions vary tremendously and range from hot, dry desert areas to mountainous regions at high elevations. The native plants that occur naturally in these areas are very different and so are the plant choices for a pollinator garden located within each locale.

Refer to Regional Plant Lists

As a bonus to this book, pollinator plant lists by region are provided at www.ecobeneficial.com/PVG. Each regional list covers several states, or one large state in the case of California. Use your regional list as a starting point when selecting plants; there will be variations in species within a given region depending on the geographical conditions of the ecoregion. If you are interested in the nitty-gritty of ecoregions, visit the EPA website on Ecoregions at www.epa.gov/eco-research/ecoregions.

Some plant groups make the list for every region in the United States. Milkweeds (*Asclepias* species) are found in each region, with approximately eighty native species in the United States. As with any plant genus, not every milkweed species will work in your garden. Your best milkweed choices depend on your location and landscape conditions.

In California alone, fifteen native species of milkweed with varying geographic adaptations are found. *Asclepias erosa* (desert milkweed) occurs only in desert areas with dry, sandy soil, while *Asclepias speciosa* (showy milkweed) is typically found in mountainous areas where soils are seasonally moist. Neither plant will succeed in the other's landscape conditions.

Plant regional milkweeds and avoid growing nonnative tropical milkweed (*Asclepias curassavica*) that can promote a parasite on monarchs, as well as reduce migration success and overall health. For an extensive explanation of this subject, please visit The Xerces Society website, www.xerces.org.

Varied flowers feed a diversity of pollinators.

Guidelines for Planting Foraging Habitat

As is the case with all planting, planting the right plant in the right place ensures gardening success. Using native plants that have evolved with the pollinators you want to attract will give your foraging habitat an advantage over gardens planted with nonnative species.

There are no hard-and-fast rules that dictate the perfect foraging habitat, nor can every site accommodate every planting goal, but below are some general guidelines.

Planting Guidelines: Foraging Habitat

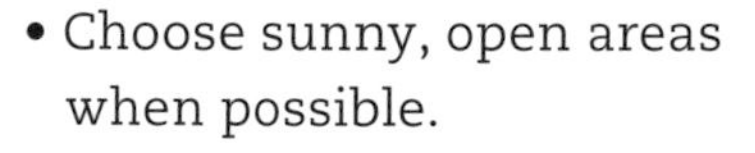

- Choose sunny, open areas when possible.
- Emphasize native plants.
- Plant for a continuous succession of bloom throughout the growing season.
- Have at least three different plant species in bloom at the same time.
- Vary flower sizes, flower colors, and flower structures of plants.
- Group each plant species into a clump 3 feet (0.28 m) square or larger.
- Space plants so they can grow to their mature width.
- Place plants close enough to eliminate large gaps between plants at maturity.
- Combine plants that have similar growth habits (e.g., clumping vs. running) and are equally competitive.

Pollinators prefer sunny gardens.

Where to Plant Forage for Pollinators

Almost any part of a landscape can be used to grow flowers for pollinators, but not all sites are equally valuable as foraging habitat. You will obviously be limited by the size of your landscape and the general conditions within it. If you are lucky enough to have parts of your landscape in full sun, you have an advantage. But even if you don't have a sunny spot, you can still plant forage plants.

Determining the Size of a Foraging Habitat

When deciding on the size of an area to plant with forage plants, go as large as you can but perhaps do so in multiple steps. If you are working with a landscaper, the work can be accomplished more easily than if you are planting the area yourself. Be realistic about your time, energy, and budget, and the scale of the project. Splitting a landscape into a series of smaller projects can be easier than doing everything at once; you can connect pollinator patches as you create them. Even if you have a very small landscape such as an urban garden or just a terrace or patio, any pollinator habitat you create is valuable, including a container garden filled with flowering native perennials.

Sunny Areas

Most forage plants, and most of their respective pollinators, prefer sunny conditions and open areas. The warmth of the sun enables cold-blooded pollinating insects to be active, and the ability to see the sky allows them to navigate. If your landscape has both sun and shade, emphasize planting most of the foraging habitat in sunny areas. An obvious area that's usually in a sunny, open location is the lawn. Convert any part of the lawn you can live without into a flower-filled pollinator buffet. It's fine to keep lawn that you really use, but lose the rest, making sure to maintain it pesticide free.

Hard-to-mow areas, like hillsides (especially if they're sunny), can be a boon to pollinators (*and* the person who has to mow) when they're converted into expanses of flowering ground covers; thicket-forming, short, flowering shrubs; or even diminutive meadows.

Shady Areas

Don't despair if your landscape is devoid of sun; you can still plant for pollinators, but your plant selections will be different. Numerous pollinator-friendly plants grow in part shade or even full shade. Shade plants can be valuable to pollinators, especially when not much else is around. In the northeastern United States, spring-blooming woodland plants including Dutchman's breeches (*Dicentra cucullaria*), trout lilly (*Erythronium americanum*), and spotted geranium (*Geranium maculatum*) are critical to early-emerging pollinators such as bumble bees and may be the only floral resources around when pollinators need them. Even summer-blooming plants like Joe Pye weed and fall-blooming shade plants can have value to hungry pollinators. Woodland asters and goldenrod (*Eurybia divariacta*, *Symphyotrichum cordifolium*, *Solidago caesia*, *Solidago flexicaulis*, and others) receive many pollinator visitors in fall.

Sunny gardens attract many pollinators.

Shady gardens can attract pollinators, too.

Diverse plantings increase the number of pollinator species you can attract.

Flower Power in the Pollinator Victory Garden

Flowers are the most important food sources for pollinators, but there's more to it than planting flowers and waiting for pollinators to appear. A successful Pollinator Victory Garden achieves a balance between two seemingly disparate principles: plant diversity and plant sufficiency. This can be thought of as "achieving floral balance." Your garden needs to have a wide variety of plant species in place to meet the varying needs of different pollinators throughout the growing season. At the same time, there must be enough of each species to entice and feed all of the pollinators that use them. It can be tricky to figure out this balance, especially in smaller landscapes. There is no perfect formula, but by keeping these concepts in mind and knowing which regional pollinators could be attracted to your landscape, you can make your garden a valuable resource.

If you keep European honey bees or if managed hives are located nearby, there will be increased competition with native bees for floral resources. While not all bee species use the same flowers, generalist foragers will use many of the same plants. In the presence of large colonies of honey bees, ramp up your floral resources, meaning: plant *a lot* more flowering plants. Plant sufficiency is also especially important in developed neighborhoods where fewer landscapes are planted for pollinators.

Boosting Plant Diversity

Diversity is critical to healthy ecosystems. Scientists have learned that landscapes planted with many different plant species are more resistant to pests and diseases and to the effects of climate change. We face all three of those significant challenges in our landscapes. The traditional gardening practice of massing huge swaths of the same species of plant may look dramatic, but from an ecological perspective, it is unwise. Balance is key when massing plants for pollinators or we lose diversity.

By planting many different species of plants in a landscape, you'll have a better chance of attracting a greater diversity of pollinators. With an estimated 200,000 species of pollinators around the world, including 20,000 bee species, any landscape planted in an ecologically sound manner may be able to attract and support hundreds of different pollinator species and thousands of different pollinators.

Not every pollinator is attracted to or can use the same species of plants. A long-tongued bee can access nectar from a long tubular flower, such as a larkspur, but a short-tongued bee cannot. Bees with shorter tongues appreciate more open flowers in a garden, particularly flowers with small nectaries (where the nectar lies within a flower). Generalist foragers can use a much greater variety of plants than specialist foragers, so plant for both. Our pollinator gardens should feed many different pollinator species, and that requires having plant diversity.

Floral targets help pollinators forage (*Coreopsis verticillata*).

Achieving Plant Sufficiency

There is a counterpoint to planting diversely. To successfully attract pollinators, you must also plant sufficiently, making it easy for pollinators to find the flowers you are offering and to provide enough of that particular resource. Plant sufficiency, or enough floral biomass, is crucial to feeding all of the various pollinators that may visit your landscape. Remember, if honey bees are nearby, their large colonies require a lot of floral food to survive.

The behavior of floral constancy in many bees and butterflies underscores the need for planting enough plants of the same species. These pollinators forage many times a day, but each time they forage, they look for one species of plant. Planting a single specimen of a perennial will not do the job for them. Think like a bee or butterfly when you go plant shopping; a pollinator needs a full meal, not just a few morsels.

There are two basic gardening approaches to provide plant sufficiency. One is to create pollinator targets and the other is to create plant repetition. Both approaches can be used in most landscapes except for very small spaces, such as a balcony or a diminutive patio garden.

Planting pollinator targets

Pollinator targets are groupings of the same species of plant, which allow pollinators to find the nectar and pollen they need easily. The achievable size for each target in your garden will depend on the size of your landscape. In a landscape with plenty of room for many different species, a target 3 feet (0.28 m) square of the same plant species, for every species, may be a reasonable goal. Realistically, not every garden has the room to achieve this

Plant repetition of a meadowscape

goal. There will always be a trade-off between planting diversely and planting sufficiently, so just do the best you can with the space you have.

While most individual trees and shrubs offer significant pollinator targets when in bloom, the majority of perennials need some companions. When shopping for flowering perennials, choose at least three plants of the same species. A single, lonely plant of small size can get lost in the garden and will be harder for a pollinator to find. If you don't have the space to include sizeable targets for each plant species, consider repeating a plant throughout your landscape.

Incorporating plant repetition

Plant repetition is a technique that can be used with pollinator targets or as a stand-alone method in a landscape. With this approach, a single plant species is repeated throughout the garden no matter the size of the landscape. A European honey bee might make fifty foraging trips in a day. She will happily find the repeating plants throughout your garden.

Not every pollinator finds flowers the same way. A compromise approach in smaller gardens is to repeat small groupings of the same plant species throughout the landscape, ensuring that all pollinators can find what they need.

One method of plant repetition, which is particularly useful where browsing deer are a problem, is to create a meadow or a meadowscape (a meadowlike garden). Natural meadows have irregular, recurring patterns of flowers, offering pollinators nectar and pollen while eliminating the easy "deer buffet" that many of our gardens have become. Learn more about meadowscapes in chapter 5.

Red maple flowers (*Acer rubrum*) are early sources of forage.

Planning for Succession and Overlap of Bloom

It's very common to see gardens that have a flush of bloom at a certain time or two of the year and then offer few, if any, flowers for pollinators during the rest of the growing season. Think of all the pollinators that go hungry when nothing is in bloom! Plant for a succession of bloom throughout the growing season and your pollinators will thank you. Include a variety of blooms that appeal to different types of pollinators to feed many different species and include plants with overlapping bloom times to ensure nectar is available.

The first step toward achieving these goals is to record the plants you already have in your garden, and when they bloom. Take a bloom inventory of flowering trees, shrubs, and perennials (but do not include any double-flowered plants). If you cannot do this from memory, fill in the blanks as flowers appear. You can be as specific with days as you like or simplify the process by using time ranges such as early spring, midspring, late spring, and so on.

To create a bloom inventory chart, record all flowering plants in your garden, as in the example below.

Sample Bloom Inventory Chart

EARLY SPRING	MIDSPRING	LATE SPRING
red maple pussy willow vernal witch hazel serviceberry redbud	golden Alexanders creeping phlox spotted geranium Virginia waterleaf wild bleeding heart	winterberry blue wild indigo spiderwort Canada violet black cherry
EARLY SUMMER	**MIDSUMMER**	**LATE SUMMER**
sundial lupine wild quinine beardtongue smooth hydrangea button bush	nodding onion common milkweed coneflower Joe Pye weed wild bergamot	purple giant hyssop rattlesnake master false sunflower obedient plant New Jersey tea
	FALL	
false aster common boneset white wood aster	flat-top goldenrod closed bottle gentian woodland sunflower	sneezeweed zig-zag goldenrod New York aster

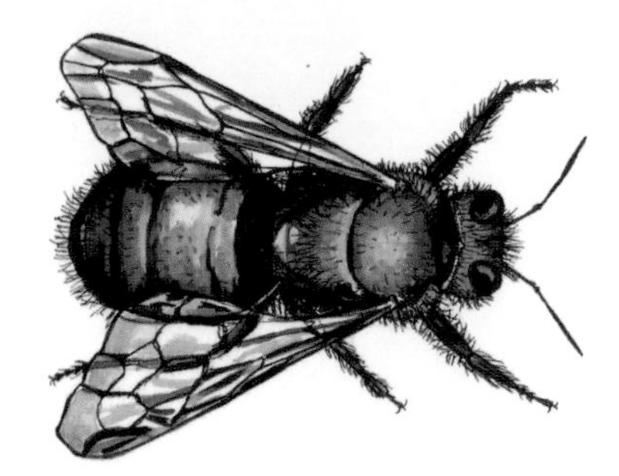

By taking a bloom inventory, you will be able to determine when the gaps in flowering occur or when there are times of minimal bloom; these are the critical times to plant for. Create an ongoing pollinator buffet filled with choices for all types of pollinators.

Very early-blooming plants are not always considered in a pollinator garden, but they are important. Bumble bees and honey bees can forage in chillier weather than some other bee species and can be active very early in the spring. Make sure to have early-flowering native plants when these insects emerge from a cold winter. A red maple blooming in March or a serviceberry blooming in April may be early feasts for honey bees and other early bees, while squirrel corn or Dutchman's breeches (*Dicentra* species) may be critical early foods for a bumble bee queen who needs to establish her new colony.

Continue the flower show until the end of the growing season, which for many regions is late fall. Quite a few pollinators are still active at that time and benefit from native plants like goldenrod and asters. Late-season blooms are often forgotten in gardens. Add them to your Pollinator Victory Garden to feed mated female bees preparing to overwinter, monarchs setting off for their migration, and hummingbirds fueling up before their long flight south.

When choosing plants, consider that not all plants have the same duration of bloom. But don't just plant those species that seem to have an endless bloom. In early spring, a serviceberry (*Amelanchier* species) may bloom for just a week or so, or even less if it rains a lot. In contrast, a purple coneflower (*Echinacea purpurea*) has a very long period of bloom, often lasting for many weeks. This,

Serviceberries (*Amelanchier* species) offer early blooms for pollinators.

however, does not make the serviceberry a bad pollinator plant compared to a coneflower. The serviceberry flowers early in the season when not much else is around to provide nectar. It's an important food source for early emerging mining bees and sweat bees as well as a larval host plant. Duration of bloom is not the sole factor when choosing pollinator plants.

It can take some time to establish a consistent succession of bloom in a pollinator garden, but your observation and planting will pay off.

Scarlet bee balm (*Monarda didyma*), a hummingbird favorite

Planting to Attract Different Types of Pollinators

Pollinators and plants have coevolved so that both parties benefit. Through evolution, pollinators adapt to certain flower traits and vice versa, ensuring that the pollinator gets the food it needs while the plant is successfully pollinated and can reproduce. These evolutionary flower traits are called pollination syndromes. Flower color, aroma, structure, and amount of nectar, the location and quality of pollen, and the presence or absence of nectar guides are all traits that determine which pollinator is likely to use a given flower.

Exceptions to these pollination syndromes do occur, especially in the absence of choice for a pollinator. Although hummingbirds prefer red, deep tubular flowers, such as those of scarlet bee balm (*Monarda didyma*), they will also use flowers without these characteristics. But why not make it easy for them and give them the flowers they prefer? While you may be tempted to purchase a yellow cultivar of coral honeysuckle (*Lonicera sempervirens*), hummingbirds may be happier with the straight species, which sports orange-red tubular flowers accented by yellow throats. Knowing the flower traits each type of pollinator has evolved with will help you make better plant choices and create a more compelling pollinator garden.

Although preferences for some flower characteristics do overlap from pollinator to pollinator, others do not. A strong, sweet fragrance emanating from a flower will not entice birds, which have no sense of smell. If that fragrance occurs at night, it will often be irresistible to nocturnal moths looking for a meal. A fresh and mild floral scent will appeal to bees, while it often takes a foul-smelling flower to attract pollinating flies. Nectar guides on flower petals will attract both bees and butterflies but they won't be noticed by a pollinating beetle. Consider these traits (pollinator syndromes) when selecting plants for pollinators.

Remember that the more flower traits you include in your garden, the wider the array of pollinators you will attract.

Flower beetle eating pollen

Pollination Syndromes (Flower Traits by Pollinator Type)

Pollination syndromes are predictive but not absolute.

Bee-pollinated flowers

Colors:	bright white, yellow, blue, violet, purple, or ultraviolet
Structure:	varied
Aroma:	mild, fragrant
Nectar guides:	present
Nectar:	usually present
Pollen:	often sticky and scented

Butterfly-pollinated flowers

Colors:	usually bright; often red, orange, yellow, or purple
Structure:	often with wide landing pad
Aroma:	slight
Nectar guides:	usually present
Nectar:	lots of nectar, deep within
Pollen:	limited

Moth-pollinated flowers

Colors:	often pale, white or pink, dull red, or purple
Structure:	clusters, landing platforms
Aroma:	strong and sweet at night
Nectar guides:	none
Nectar:	lots of dilute nectar, deep within
Pollen:	limited

Bird-pollinated flowers

Colors:	scarlet, red, orange, or white
Structure:	large, funnel-shaped
Aroma:	none
Nectar guides:	none
Nectar:	lots of nectar, deep within
Pollen:	some

Bat-pollinated flowers

Colors:	often white or pale, green, or purple
Structure:	usually open at night
Aroma:	highly fragrant, fruity or fermenting
Nectar guides:	none
Nectar:	lots of dilute nectar
Pollen:	lots

Beetle-pollinated flowers

Colors:	often dull white or green
Structure:	often bowl shaped
Aroma:	strong, fruity or fetid
Nectar guides:	none
Nectar:	usually present
Pollen:	lots

Fly-pollinated flowers

Colors:	often dark brown, purple, or pale
Structure:	often funnel shaped or complex
Aroma:	putrid, rotting flesh smell
Nectar guides:	none
Nectar:	usually absent
Pollen:	some

Food for Butterflies and Moths: Larval Host Plants

Flowers are arguably the best known and most important source of food for pollinators, but for butterflies and moths, nonfloral foods called larval host plants are critically important and cannot be left out of a discussion about food for pollinators. Butterflies and moths undergo complete metamorphosis, meaning they have four distinct life stages: egg, caterpillar (larva), pupa (chrysalis), and adult (imago). While the majority of adult butterflies and moths eat nectar, the preponderance of their caterpillars have a completely different diet.

When planting a garden, we may not have considered what caterpillars eat, but it is extremely important for supporting butterflies and moths. Most caterpillars eat plant parts, typically leaves from larval host plants, specific plants with which the caterpillars have evolved. Some caterpillar species eat flower buds and flower parts, fruits, seeds, other caterpillars, or even animal feces. But by and large, caterpillars eat leaves of their larval host plants.

Types of host plants

Host plants can be trees, shrubs, vines, perennials, grasses, or sedges—it depends on the caterpillar species in question. Trees and shrubs are particularly important host plants—many, but not all, native woody plants serve as larval hosts. Some species can sustain hundreds of different caterpillar species. Research conducted by Dr. Douglas Tallamy, author and professor at the University of Delaware, concluded that in the mid-Atlantic region of the United States, native oaks as host plants support more caterpillar species than any other trees or shrubs: *557 total caterpillar species*. Native cherries and willows are close behind, feeding 456 and 455 caterpillar species respectively.

Although it may be easy to see where adult butterflies are nectaring, it can be much harder to discern where their caterpillars are feeding. Since most moths are nocturnal feeders, their larval food plants can be even more difficult to determine by observation. Your best tools to determine the host plants that feed specific caterpillars are in-depth books and credible websites on caterpillars, butterflies, and moths. Also check out the resources in the appendix of this book and the

Spicebush swallowtail butterfly caterpillar eating the leaves of one of its host plants, spicebush (*Lindera benzoin*)

list of common host plants and the caterpillars they feed, available at www.ecobeneficial.com/PVG.

Adding host plants

If you have a very small garden without enough room to plant a variety of host plants, you can "borrow" or "share" host plants with neighbors. Talk with your neighbors to form a plan to fill in the host plant gaps in order to attract more butterflies and moths.

To make sure that you have the right larval host plants available, create a host plant checklist for your landscape. The first step is to determine the butterfly and moth species that are common to your area. There are online and printed resources to help you (see the appendix). Set up a simple checklist to note the host plants you already have in your garden that regional butterflies and moths look for. Don't have enough host plants in your garden? Then it's time to plant more.

(spp. = species) **Sample Host Plant Checklist**

BUTTERFLY	HOST PLANTS	IN MY GARDEN?
eastern tiger swallowtail	***Betula*** spp. (birches)	☐
	Fraxinus spp. (ash)	☐
	Liriodendron tulipifera (tulip tree)	☐
	Magnolia virginiana (sweetbay magnolia)	☐
	Populus spp. (cottonwoods, poplars)	☐
	Prunus spp. (cherries, plum)	☐
	Salix spp. (willows)	☐
	Tilia americana (basswood)	☐
red spotted purple	***Betula*** spp. (birches)	☐
	Crataegus spp. (hawthorns)	☐
	Populus spp. (cottonwoods, poplars)	☐
	Prunus spp. (cherries, plum)	☐
	Salix spp. (willows)	☐
	Quercus spp. (oaks)	☐

From the sample checklist above, you could determine whether your existing garden can support either butterfly in question. By conducting this brief exercise, you would also discover that several plants (in bold) will feed multiple species of caterpillars found in your region that you want to support. This preplanting exercise can save you time, money, and planting space *and* feed more types of caterpillars.

Host plants and evolutionary connections

Adult moths and butterflies have evolved to instinctively sense where to lay their eggs (oviposit) on or near the larval host plants appropriate for their particular species. This is one reason native plants are so important in a pollinator garden, because native pollinators and native plants have evolved together. When the eggs hatch, caterpillars first eat their own eggshells and then have an immediate food supply nearby from their host plant. While some butterflies and moths have evolved to be able to depend upon a wide range of larval host plants, others have a very limited diet; the monarch may be the best-known example of the latter.

Monarch caterpillars eat the leaves of milkweed, their host plants (*Asclepias* species)

Monarch butterflies have evolved to use plants in the milkweed group (the genus *Asclepias*) as host plants. Although adult monarchs will sip nectar from many different flower species, including milkweeds, their caterpillars eat *only* milkweed leaves. If you want to support monarchs, which have declined by at least 70 percent since their population high in the 1990s, include milkweeds in your pollinator habitat. Native milkweeds are also excellent nectar plants for a variety of pollinators.

Tolerating plant damage

For many homeowners, the idea of encouraging the chewing of plant leaves may be a bit hard to accept, but that is what most caterpillars do. Sometimes the leaf chewing by caterpillars appears quite minor, while at other times it may be more extensive. Caterpillars must eat a lot to grow; a newly hatched caterpillar may increase its body mass 10,000-fold in just two weeks. If you want butterflies and moths, accepting leaf damage is part of the equation. It will be worth it when you see adult butterflies flourishing in your landscape.

A proud client once pointed out her leafless milkweed plants happily exclaiming that she had raised an abundance of monarchs that year with the milkweed we had planted in her garden. Further evidence was the bevy of adult monarch butterflies busily nectaring on other flowering plants around us. Becoming a pollinator steward means expanding the definition of what a garden can be, which is a healthy ecosystem where some plant damage is perfectly acceptable for the greater good.

Redbud (*Cercis canadensis*), an early pollinator favorite

Adding Biomass with Trees and Shrubs

While flowering perennials are typically the mainstay of a pollinator habitat, don't forget their woody friends: flowering trees and shrubs. These woody wonders have a particular appeal as pollinator plants with an abundance of blooms on a single plant. Flowering trees and shrubs of size can ramp up your pollinator foraging resources, which is particularly valuable for large colonies of honey bees. Trees, shrubs, and vines are often overlooked as pollinator plants, and that's a shame. For example, red maple (*Acer rubrum*) flowers are one of the earliest sources of nectar and pollen for bees in the northeastern United States, making that tree a valuable addition to a pollinator garden in that region.

Small trees and shrubs can be used as foundation plantings against a house, combined diversely as a hedgerow for pollinators, or planted wherever they fit into your landscape. They can even be planted around the edges of a dedicated pollinator garden as long as they don't shade out the open, sunny areas. Determine the mature height and width of a plant *before* you plant it so you don't create a shade garden later.

Incorporating Nonfloral Foods for Pollinators

Not all pollinators get their complete nutrition from flowers. A number of nonfloral foods are eaten by pollinators, depending on the pollinator species. In the case of butterflies, mourning cloaks, question marks, eastern commas, and red admirals favor tree sap, rotting fruit, aphid honeydew (aphids' liquid excrement), bird droppings, and animal feces. Even carrion can be enticing to some butterflies. In North America, there is one type of carnivorous butterfly, the harvester, that specializes in eating woolly aphids.

Hummingbirds mostly eat nectar, but they also feed on tree sap and juices from overripe fruits. For protein, they eat pollen and small insects. Pollinating bats feed on insects as well as nectar, pollen, and flower parts, depending on the bat species.

Providing these alternative food sources is best done by maintaining a healthy and well-balanced ecosystem. Include a few sap-producing trees, some fruit-bearing trees and shrubs, and quality habitat for birds (bird droppings!), and avoid the use of pesticides. In a pollinator garden, aphids actually have value, because they produce honeydew for butterflies and a protein snack for lady beetle larvae (which, conversely, keep aphid populations in check).

A Man-Made Butterfly Buffet

If you want to supplement natural foods, you can create a butterfly buffet of rotting fruit. Fill a plate with overripe fruit and suspend it from a tree with thin cord or fishing line to protect it from hungry mammals. Leave it out only on sunny days when butterflies are active or leave it out at night for nocturnal moths to feed on. Just place the dish where you can easily clean and replenish it.

Hummingbird Feeders

Hummingbirds appreciate a hummingbird feeder in addition to flower nectar, if the contents of the feeder are replaced frequently and *the feeder is kept clean*. Glass feeders are easier than plastic ones to clean thoroughly. Skip the red dye, which may not be safe, and stick to the standard formula of one part sugar to four parts water, boiled together to form a sugar mixture. You'll also save hummingbirds from expending valuable energy if you provide a feeder that has places to perch while the hummingbird nectars.

Puddling

In nature, butterflies often gather at moist, sandy, or gravelly spots and use their tongues to suck up moisture containing salts and minerals. This behavior, called puddling, is mostly seen in male butterflies. There is speculation as to why butterflies do this; it may be that they need additional sources of sodium and minerals beyond what they get in nectar.

This sodium deficiency may be more severe in male butterflies during mating, potentially explaining why the majority of puddlers are male. It has also been speculated that males use puddling as a way to incorporate nutrients into their sperm, which is then transferred to the female during mating, as a way of increasing egg viability.

Although it is not clear what puddling behavior really represents, it seems to be important, at least for males. Notably, butterflies also feed on animal dung, urine, sweat, and carrion, which are possibly other ways to obtain sodium.

You can create a puddling habitat in your landscape by filling a large clay saucer with coarse sand, gravel, and/or mud and topping it off with water to make a damp mix. Add a dash of finished leaf or manure compost along with a sprinkle of salt to boost the nutrients. Keep the mixture consistently moist and occasionally replenish the added nutrients. Don't be disappointed if butterflies don't show up; it can be hit-or-miss with a homemade puddle. The

Butterflies puddling

easiest approach is to leave existing puddling sites undisturbed, perhaps adding moisture when needed to prevent them from drying up.

Insect Meals

Some pollinators, including hummingbirds, bats, and wasps, have a varied diet that includes insects. A healthy pollinator garden is filled with many different types of insects, which are an important source of protein. Sometimes pollinators wind up being that protein source for a hungry bird. Just remember your garden is an interconnected food web, and all creatures have to eat.

4 Parade of Pollinators

Animals that pollinate flowers include creatures that are very familiar, such as bumble bees and honey bees, as well as less obvious animals. There are pollinating beetles, bats, butterflies, moths, lizards, rodents, lemurs, honey possums, and even monkeys. Not every species within a group of animals is a pollinator. Approximately 9 percent of all birds and mammals on Earth are pollinators. One of the most unusual mammal pollinators is the lemur, which is the largest pollinator in the world.

Bumble bee nectaring on goldenrod (*Solidago species*)

Bloom Considerations for Bees

Since bee species vary in their shapes, sizes, body strengths, tongue lengths, and flower preferences, a diversity of flowering plants is needed to satisfy them all. And, when planting, don't forget to include plants for the 25 percent of bee species that are pollen specialists. To keep everybody happy, keep the flower show going from early spring through late fall in most regions. In subtropical or temperate desert areas, plant winter-blooming plants as well.

Bees create nests which must be provisioned with nectar and pollen to feed their brood. Some bees store nectar in their nest for adult food reserves. How they provision depends on the bee species, but in every case, it helps bees save energy when floral resources are reasonably close to their nesting site. While honey bees can fly several miles (km) and bumble bees might be able to fly a mile (1.6 km), some quite tiny bees have flight ranges of just a few hundred feet (m). Flying expends energy, so the closer food is for pollinators, the less energy they have to use.

Open flowers of native asters are easily accessed by bees.

Bee-Pollinated Flower Characteristics

Flowering plants that are pollinated by bees tend to fall into two distinct categories:

- simple, open, or bowl-shaped flowers that are relatively unspecialized (example: native roses, sunflowers)
- complicated, asymmetrical flowers that are more specialized (example: beardtongues, native orchids)

Flowers in both groups are usually blue, violet, yellow, or white; often have some scent; and are open during the day. Ultraviolet nectar guides are typically present on these flowers. The nectar quality and pollen quality will vary depending on plant species. Bees often favor flowers where the nectar's sugars are dominated by sucrose. Pollen of bee-friendly flowers may be scented and sticky, and it must be shaken loose from a flower's anthers by a pollinator.

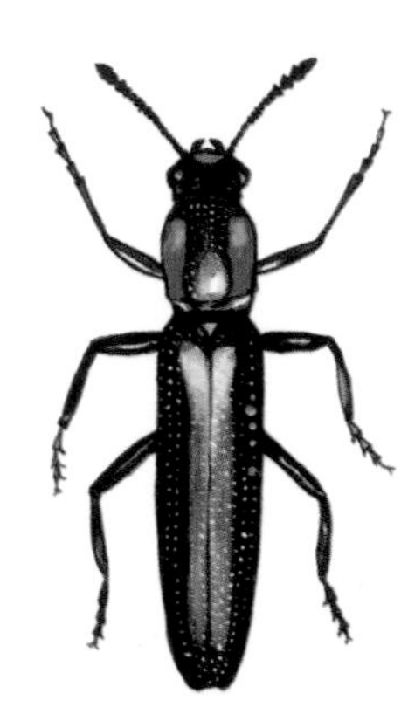

Native trees and shrubs can be major nectar sources for honey, too, including black locust (*Robinia pseudoacacia*), tulip tree (*Liriodendron tulipifera*), basswood (*Tilia americana*), sumac (*Rhus* spp.), and raspberries and blackberries (*Rubus* spp.). Mono-floral honeys from several native trees are particularly delicious and highly valued; sourwood (*Oxydrendron arboreum*) and tupelo (*Nyssa sylvatica*) honeys are two notable examples.

Early-spring-blooming plants such as red maple (*Acer rubrum*) and willow (*Salix* spp.) may be good forage resources for honey bees when little else is in bloom. If weather conditions are warm enough, honey bees benefit from these plants. If the weather is cold, they may miss these blooms.

The flower traits that attract native bees also attract honey bees, with the main difference being flower anatomy. Honey bees have easiest access to more open flowers with simple structures. As with native bees, honey bees benefit from a varied diet with different sources of nectar and pollen.

Floral resources for honey bees often diminish when there are extended periods of hot and dry weather and nectar stops flowing. These periods are known as summer nectar dearths. However, nectar dearths can occur at other times of the year as well. By planting more generalist plants, especially ones that flower in summer, you can help honey bees find the food they need during difficult times. If you keep bees, you may want to include some well-behaved nonnative plants in your landscape for additional forage.

Helping Honey Bees

You don't have to keep beehives in order to help honey bees. A Pollinator Victory Garden can provide support to managed honey bees and feral honey bees as well as native bees. Include some of the native plants that are accessible to honey bees. Trees and shrubs offer the most floral mass for large honey bee colonies. Add more bloomers for that critical time of summer dearth. And, of course, keep your garden free of pesticides. If you live near beekeepers, they will greatly appreciate the floral resources you offer in your garden. You might even be rewarded with some honey!

Honey bee nectaring on purple coneflower (*Echinacea purpurea*)

Examples of Native Flowers Pollinated by Honey Bees

Agastache foeniculum (anise hyssop)
Chamerion angustifolium (fireweed)
Echinacea spp. (coneflowers)
Eriogonum spp. (buckwheats)
Eutrochium spp. (Joe Pye weeds)
Helianthus spp. (sunflowers)
Phacelia spp. (phacelias, scorpionweeds)
Rhus spp. (sumacs)
Solidago spp. (goldenrods)
Symphyotrichum spp. (asters)
Vernonia spp. (ironweeds)
Veronocastrum virginicum (Culver's root)

Eastern tiger swallowtail nectaring on Turk's cap lily (*Lilium superbum*)

Butterflies

Butterflies are perhaps the most charismatic visitors to our gardens. It can be easier to convince someone to plant a butterfly garden than a bee garden, but, fortunately, both groups use many of the same plants. Of the 20,000 species of butterflies in the world, 725 are found in North America, with 575 of those species in the forty-eight contiguous states.

Butterflies are often colorful and easy to recognize, while others, specifically skippers, are frequently small, brown, and mothlike. Skippers are a subset of butterflies, accounting for about 270 of the 575 butterfly species in the continental United States. Butterflies cannot always be easily distinguished from moths, although butterflies tend to rest with their wings closed or slightly open in a V shape while moths tend to hold their wings open when at rest. Most butterflies have antennae that are club-shaped at the end, while most moths do not.

Resident and Migratory Species

Many butterfly species are local and stay local, but others migrate. Migrating butterflies that travel to some degree include painted ladies, red admirals,

Silver-spotted skipper on wild bergamot (*Monarda fistulosa*)

cloudless sulphurs, mourning cloaks, and question marks. The butterfly champion of long-distance migration is the monarch, which can travel extraordinary distances to escape cold winters.

There are two migratory populations of monarchs in North America. Those that live west of the Rocky Mountains overwinter in Southern California along the Pacific Coast. This western population is critically endangered. Monarchs that frequent gardens in the eastern United States have evolved to migrate in the fall to Mexico, where they overwinter in a small tract of forest, which, sadly, is diminishing.

The eastern monarch population has declined by approximately 70 percent since its recorded high in the 1990s. The same individual monarchs do not travel back and forth from the east to Mexico. The first generation of monarchs are the offspring of those that have overwintered in Mexico. It takes three to four generations to travel to the northern United States and Canada. Plant milkweeds, monarchs' obligate host plants, to help them survive these incredible odds. Once very common, the monarch is now threatened with extinction. In addition, more than twenty other species of butterflies are at risk of extinction in the United States, and many more are threatened. Get busy planting for them!

Moths

Compared to their butterfly cousins, moths are the less-appreciated members of the insect order *Lepidoptera*, which includes butterflies and moths. While butterflies are active during the day, most moth species nectar at night, although some adult moths eat nothing at all. The enchanting hummingbird moth is diurnal (active during the day). This creature has the perfect common name—it has a mothlike body and flies like a hummingbird.

Many more moth species exist in the world than butterfly species—about 160,000 globally, with approximately 11,000 species in the United States. Moths are most reliably distinguished from butterflies by the placement of their wings while at rest; they hold their wings open. Although moths often have duller coloration than butterflies, there are some pretty spectacular moth species, like the cecropia, luna, and white-lined sphinx moths.

Moth Habitat and Life Cycle

Like their butterfly cousins, most moths are solitary and do not have nests. Like butterflies, moths also need larval host plants on which to lay their eggs. The caterpillars that emerge from the eggs eat the larval host

A bird-dropping moth enjoying the nectar of silky prairie clover (*Dalea villosa*)

plant's leaves until the caterpillars get large enough to realize it's time for their next life stage. They then leave the host plant and search for a place to reach their next life cycle. Some species of moth caterpillars burrow into soil or hide under dead leaves to pupate; other species form a cocoon, which they may attach to a plant.

Flower longhorn beetle eating pollen

Beetles

Beetles are thought to be some of the most ancient pollinators. They began visiting flowering plants 150 to 200 million years ago, much earlier than bees. Beetles are also the largest group of animals on Earth, with an estimated 350,000 species globally, of which approximately 30,000 species occur in the United States. However, not all beetles are pollinators. Although few plants depend solely on beetle pollination, beetles are important pollinators of ancient plant species such as magnolias (*Magnolia* spp.), spicebushes (*Lindera* spp.), and water lilies (*Nymphaea* spp.).

Beetles move slowly and cannot visit as many flowers as some other pollinators, making them less effective than other pollinator groups. Even so, pollinating beetles have hairy bodies that catch pollen, and they manage to get the pollination job done. Some beetle species feed on both nectar and pollen, but others feed only on pollen, so the plants that attract them typically have lots of pollen and usually some nectar (although no nectar guides are present). Beetles may also eat flower petals and other flower parts.

Common beetle pollinators include flower long-horn beetles, soldier beetles, leaf beetles, pollen beetles, and tumbling flower beetles. The habitat that pollinating beetles need matches the habitat you create for bees and butterflies.

Beetle-Pollinated Flowers

Although beetles see colors, they rely on their sense of smell to find flowers. Beetle-pollinated flowers are often dull white or green, and based on the flowers they visit, maroon appears to be attractive to them as well. Beetles nectar from flowers of other colors if they are drawn by the scent. Strong, fruity, or fetid odors appeal to beetles. Open flowers and bowl-shaped flowers are the easiest for beetles to access.

Examples of Native Flowers Pollinated by Beetles

- *Allium* spp. (native onions, garlics, ramps)
- *Arisaema triphyllum* (Jack-in-the-pulpit)
- *Aruncus dioicus* (goat's beard)
- *Asclepias* spp. (milkweeds)
- *Asimina* spp. (pawpaws)
- *Calycanthus* spp. (sweetshrubs)
- *Eryngium* spp. (eryngos, rattlesnake master)
- *Illicium floridanum* (Florida anisetree)
- *Lindera* spp. (spicebushes)
- *Magnolia* spp. (magnolias)
- *Malus* spp. (crabapples)
- *Nymphaea* spp. (water lilies)
- *Solidago* spp. (goldenrods)
- *Spiraea* spp. (spiraeas, meadowsweets)
- *Symphyotrichum* spp. (asters)

Great black digger wasp foraging on mountain mint (*Pycnanthemum* species)

Wasps

Not many people get excited when they hear about gardening to support wasps. The truth is, wasps are valuable insects in the landscape, particularly as predators and parasitizers of insect pests. Bees have descended from wasps and are in the same order of insects (*Hymenoptera*), so wasps and bees have similar life cycles and nesting behaviors.

Wasps often look like bees but usually don't have the same fuzzy hairs on their bodies that enable bees to pick up pollen. This means that less pollen sticks to their bodies, limiting their effectiveness as pollinators.

Solitary vs. Social Wasps

Social wasps are the insects that people usually think of when it comes to wasps, such as paper wasps and yellow jackets. These insects nest in colonies, some in the ground and some in structures they build. Solitary wasps make up the preponderance of wasps and nest alone in the ground, in pithy plant stems, or in structures they create with mud. Both social wasps and solitary wasps visit flowers for nectar.

Parasitic wasps (parasitoids or parasitizers) are solitary and have a fascinating lifestyle, in a gruesome sort of way. They lay their eggs in or near the bodies of other insects. Once the wasp larvae emerge, they eat the body of their insect host. Parasitic wasps are important beneficial insects with a well-earned reputation as garden allies.

Wasp-Pollinated Flowers

Wasps require a habitat that has areas to nest in and an abundance of insects to feed their young. They are terrific predators; during their larval stage, wasps are carnivorous, eating insect prey collected by their mother. When wasps become adults, they visit flowers for nectar, but some are also carnivorous. Wasps have short tongues and need more open, accessible flowers than bees need. They are attracted to many of the same flowers that attract butterflies.

Examples of Native Flowers Pollinated by Wasps

Arnoglossum atriplicifolia (pale Indian plantain)
Asclepias spp. (milkweeds)
Eryngium spp. (eryngos, rattlesnake master)
Eupatorium perfoliatum (common boneset)
Helianthus spp. (sunflowers)
Heliopsis spp. (false sunflowers)
Parthenium integrifolium (wild quinine)
Pycnanthemum spp. (mountain mints)
Solidago spp. (goldenrods)
Symphyotrichum spp. (asters)
Zizia spp. (Alexanders)

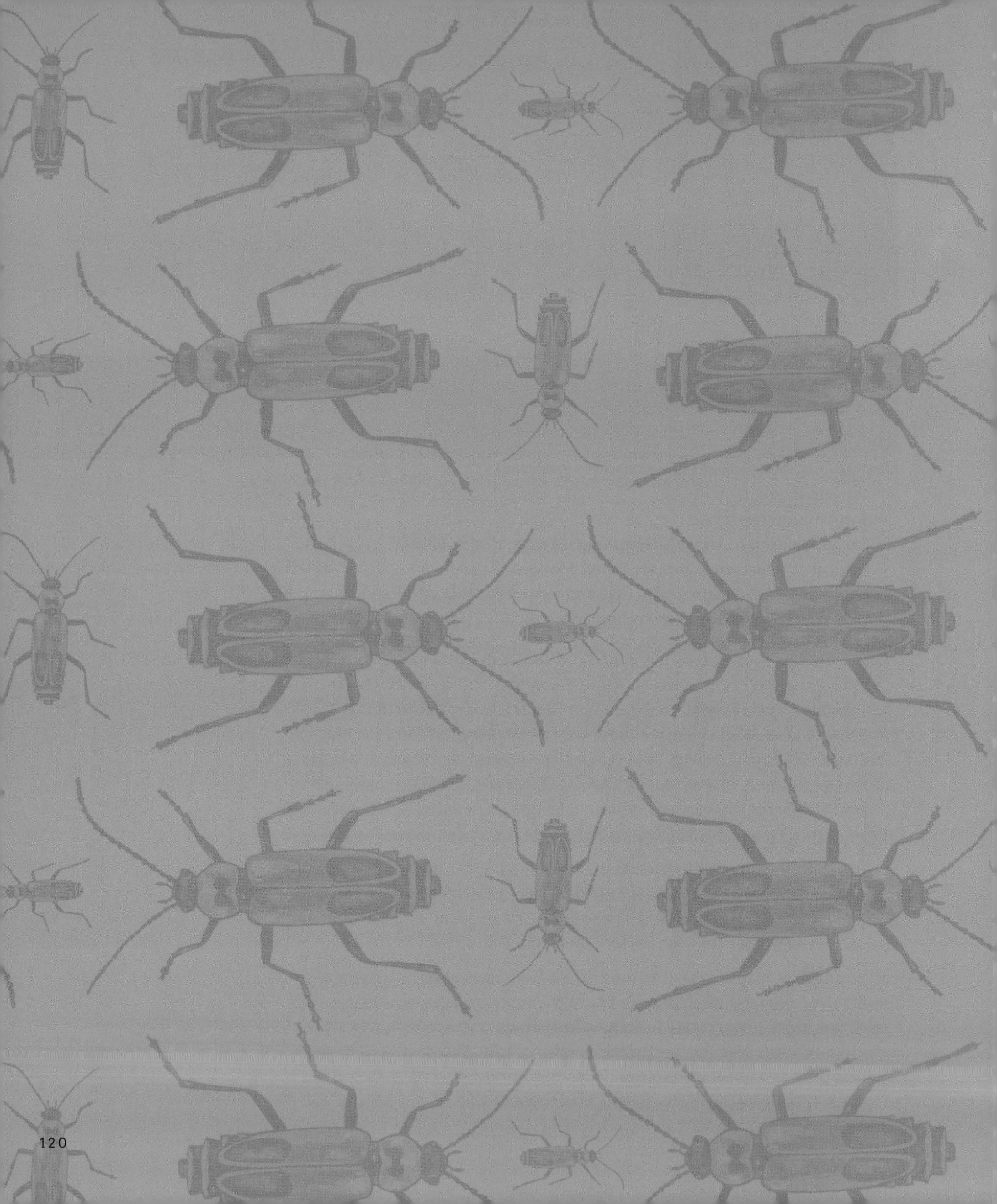

5 Creating and Growing a Pollinator Victory Garden

Now that you know the types of pollinators you can expect to see, the types of habitat to provide, and the types of forage to offer, it's time to start creating your Pollinator Victory Garden.

A Pollinator Victory Garden can take many forms. The size and conditions of your landscape, your available time, and your budget will be the major factors determining the type of garden you create. If you have an existing flower garden, you might start simply by tweaking that space to include more pollinator plants and nesting habitat. But if you want to really devote your landscape to supporting pollinators, more detailed updates may be in order.

Planting pollinator islands

Pollinator Victory Garden Projects

Getting started with pollinator gardening can be intimidating. It can be hard to decide where to plant, what to plant, how much to plant, and so forth. If a planting project is overwhelming, even the most avid gardeners, regardless of their good intentions, may simply put it off.

One solution is to start with a small project that can easily be envisioned, planted, and appreciated within a reasonable time frame. Here are some Pollinator Victory Garden project ideas that are achievable and will attract an abundance of pollinators to your landscape.

Consider under-used pollinator plants, such as prickly pear cactus (*Opuntia humifusa*) for hot, dry pollinator islands.

Pollinator Islands

Imagine creating a pollinator haven in the middle of a pollinator wasteland! Pollinator islands are a great way to support pollinators while reducing the amount of lawn ("green desert") on your property. Since removing and replacing a lawn can be tedious, begin with a small area that seems manageable.

Find the perfect location

Select a location for your first pollinator island that you can easily see. Once the pollinators appear, you will likely be inspired to create even more pollinator islands. Consider a location in the front yard, which may help to inspire neighbors, passersby, and even the mail carrier to try pollinator gardening at their own homes.

Define the area to be planted

Make the pollinator island large enough to contain a diversity of plants, making sure to leave enough room to have a quantity of each species. Grouping each species together will provide a more traditional look and create better pollinator targets.

An irregular shape with serpentine curves is always pleasing to the eye and resonates with onlookers as it feels like an intentional garden. Since the best practice for any ecological garden is to leave perennials standing through winter to provide habitat for insects and seeds for wildlife, the neat delineation of a pollinator island serves as a "cue to care" and helps offset the presumed untidiness of dead perennials in winter.

The easiest way to create the shape of the island, or any new planting bed, is to use temporary chalk paint spray to mark the area. If you decide to change the shape, spray again; the remaining chalk will wash away.

Pick out meadow plants

While flowering native perennials and biennials (forbs) are the go-to plants for pollinators, native grasses are an important component of native meadows and meadowscapes. Native grasses provide a matrix aboveground and belowground, supporting and knitting together a complex, interactive community of plants. Their deep roots are also very effective at preventing soil erosion. Most meadows consist of at least 40 percent native grasses. Fortunately, many native grasses are also larval host plants and seed sources for birds and small mammals, as well as habitat for many creatures.

Plant the meadowscape with seeds or plants

Meadowscapes planted with seeds are the most cost effective, random, and naturalistic. Seeded meadows also require the most initial management and take longer to completely establish—typically several years.

Meadowscapes planted with perennial plugs or small plants are a bit more expensive, can be more intentionally designed, and can establish within a growing season or two, depending on the time of year they are planted. When designing a meadow, consider creating a path within it so you can see the pollinator action up close.

If you like the idea of a meadowscape but feel the need for a neater aesthetic, try a "meadowish" approach. Use plant species typically found in meadows (such as native penstemons, coneflowers, goldenrods, asters, grasses, and so on), but group each species into intentional swaths using live plants; think of a perennial bed that has been "meadowfied."

Convert an unusable hillside into a meadow

If you have an unusable hillside currently planted in turfgrass, consider planting a low meadowscape to help pollinators while ending the nightmare of mowing a slope. Short native perennials and grasses, such as tickseed (*Coreopsis* spp.) and prairie dropseed (*Sporobolus heterolepis*) work well and will hold soil and prevent erosion when thickly planted.

Typical meadow/prairie plants

Maintain the meadow for longevity

After establishment (once plants have all rooted and filled in), meadows and meadowscapes must be periodically cut back, typically once a year in early spring. If this cutting is not done, woody plants may soon encroach on the meadow. Keep the perennials and grasses standing through winter to provide habitat and seeds for wildlife. Meadows of all kinds are slow to start in the spring; most meadow species, especially native grasses, are warm-season plants. In spring, for larger meadows, consider cutting smaller areas back over a period of weeks, allowing hibernating species to wake up and leave without injury.

Keep nectar flowing by planting the right plant in the right place with proper moisture.

Growing and Maintaining a Pollinator Victory Garden

All newly-planted plants, even native plants, need your attention for the first year or two. When Mother Nature doesn't deliver enough rain, consider supplemental watering in the first two years to ensure that roots establish well. Once native plants are well-rooted, they will be very low maintenance. If you have planted from seed, be patient; it may take a while until seeds germinate and the plants fill in.

A seasonal calendar will help you establish a good maintenance regimen. At right are some tips to keep your garden and the pollinators healthy and happy for years to come.

Seasonal Pollinator Victory Garden Maintenance*

Spring

- Wait to cut back last year's perennials until daytime temperatures are in the mid-50s°F (12 to 13°C). Your plants may still be providing shelter to overwintering pollinators.
- Leave 6 inches (15 cm) or more of stubble as pollinator habitat when cutting back dead perennials. Leave some pithy plant stems standing as they may be sheltering cavity-nesting bees.
- Stagger your cutting back over a period of days or weeks to allow different types of pollinators to warm up and become active.
- Avoid stepping on wet soil, which will compact it and wreak havoc with plant roots. Wait until the ground has dried a bit to start working in the garden.
- Check to see what early-blooming pollinator plants are in your garden and plan to add more. Early spring can be a challenging time for pollinators.
- Do your planting while temperatures are mild and rain is sufficient.

Summer

- Keep newly planted plants properly watered to get their roots well established.
- Don't let flowering perennials droop in periods of heat and drought—nectar will not flow sufficiently.
- Note any times when there is a dearth of summer blooms. Make a plan to add more plants that flower at that time.

Fall

- Check to see what late-blooming pollinator plants are in your garden and plan to add more. Make sure that overwintering pollinators go into winter well fed.
- Fall is a great time for planting in much of the United States. Add more plants as needed to your Pollinator Victory Garden.
- Leave fallen leaves in place as habitat for pollinators. Skip the leaf blowers to protect pollinators.
- Don't cut back perennials in fall but leave them standing through winter as habitat.
- Keep native grasses standing through winter for the same purpose.

Winter

- In warmer regions where fall is too hot for planting, add new pollinator plants to your garden.
- Keep ice melt and salt away from your Pollinator Victory Garden. Many plants are intolerant of these substances.
- Start planning the next phase of your Pollinator Victory Garden.
- Research any native plant sales coming in spring.
- Help to plan a pollinator garden for your community, school, house of worship, or a nonprofit organization.
- Buy a pollinator habitat sign to install when the ground thaws.

*Adjust this calendar in warmer climates.

Pollinator Victory Garden Checklist

To get your Pollinator Victory Garden really working and to keep the momentum going, answer the following questions now and over time. This will help you plan and implement the steps needed to have a thriving pollinator garden. Share this checklist with family, neighbors, and friends who want to help pollinators too.

☐ Have you removed some of your lawn and replaced it with flowering native plants?

☐ What percentage of lawn did you replace? ____________________
Can you replace more? ________ Where and when? ____________________

☐ Name the flowering native trees, shrubs, and vines (woodies) in your landscape:

Spring bloomers ____________________
Summer bloomers ____________________
Fall bloomers ____________________
Winter bloomers (if applicable in your region) ____________________

☐ What flowering native woodies could you add, and when do they bloom?
____
____

☐ Name the flowering native perennials, biennials, annuals, and ephemerals in your garden:

Early spring bloomers ____________________
Late spring bloomers ____________________
Early summer bloomers ____________________
Late summer bloomers ____________________
Early fall bloomers ____________________
Late fall bloomers ____________________
Winter bloomers (if applicable in your region) ____________________

☐ Do you have at least 3 of each of the plants above? Which ones do you need to add more of? ____________________
____

☐ Are there any times of the season when not enough plants are in bloom? List the seasons and the plants you could get to fill those gaps:

Early spring bloomers ____________________
Late spring bloomers ____________________
Early summer bloomers ____________________
Late summer bloomers ____________________
Early fall bloomers ____________________
Late fall bloomers ____________________
Winter bloomers (if applicable in your region) ____________________

☐ Do you have a diversity of flower colors, shapes and sizes in each season? If not, what are you lacking ____________________
____

☐ What native plants could you get to fill in what is missing?
____
____